AX25

$ 11

M000308627

Phil Hall

FEARSOME FORDS 1959-73

Motorbooks International
Publishers & Wholesalers Inc
Osceola, Wisconsin 54020, USA

First published in 1982 by Motorbooks International Publishers & Wholesalers Inc, PO Box 2, 729 Prospect Avenue, Osceola, WI 54020 USA

Printed and bound in the United States of America

Library of Congress Cataloging in Publication Data

Hall, Phil
Fearsome Fords, 1959-1973.

Includes index.
1. Ford automobile. I. Title.
TL215.F7H285 629.2'222 82-7849
ISBN 0-87938-138-8 AACR2

Cover photo: 1970 428 Cobra Jet Mach I Mustang, by Jerry Heasley

Motorbooks International books are also available at discounts in bulk quantity for industrial or sales-promotional use. For details write to Special Sales Manager at the Publisher's address

ACKNOWLEDGMENTS

A third-grade school project was to write a business letter to some company requesting material that could be shown in class. I chose the Ford Motor Company and remember putting a three-cent stamp on an envelope, sending it off and getting a big package of photos and literature in the mail.

It stuck in my mind; all that for three cents! From that day on each year letters went out to the auto companies requesting photos and literature.

Doing that for nearly thirty years, plus going to dealers, auto shows, swap meets, press showings, etc., has resulted in a rather large collection of material.

The Ford portion of the collection, including sales literature, press kits, press books, Buyer's Digests, service manuals and technical publications, formed the basis for the primary sources for this book.

However, since no collector has all the material there is on any major topic, outside sources were sought to fill many of the gaps.

The project was lucky enough to receive outstanding help from Bob Costanzo of Daytona International Speedway, Leslie Lovett of the National Hot Rod Association, Bill Trainor and Bud Krause of Zecol-Lubaid, Carol McFarland of Edelbrock Equipment, Gene McKinney of the Motor Vehicle Manufacturers Association and Chris Poole of *Consumer Guide*.

Individuals who pitched in when help was needed include Ray Lemmermann of Ray-Mars Photos, literature collector Gary Mihelich, stock car builder Bill Behling, former driver Tom Pistone, King Cobra owner Steve Honnell and 1955-57 Ford performance expert Rick Reichardt.

Acknowledgments cannot be complete without special recognition of the most enthusiastic Ford performance fan I've ever met, Stuart Sternberg, who literally turned over his entire collection for the project, then went out and put in countless hours chasing down photos and information I didn't have time to search out.

Secondary sources included surveying most of the automotive magazines of the 1959-73 period, most of which are mentioned in the text. Also, current-day publications were very helpful, like Chick Schiesser, Jr.'s *Speedway Fords & Mercurys* and John Paradise's *Super Fords*.

Many books were consulted during the project and a few proved to be outstanding including Pat Ganahl's *Ford Performance*, Leo Levine's *Ford: The Dust and the Glory*, Wally Wyss's *The Super Fords*, and the *1964½ Thru 1973 Mustang Recognition Guide* by Editors of *Mustang Monthly*.

Production figures for Fords came from Jerry Heasley's *The Production Figure Book for U.S. Cars*, while other figures were used from *The Great Camaro* by Michael Lamm and *Automotive News*.

Last and not the least appreciated was the cooperation from my employer, Jim Engel of *Midwest Racing News and Wisconsin Auto Racing*, who gave me free use of the photo and reference files and even let my work schedule be rearranged so the project could stay on schedule.

Before getting into writing, I used to be amazed at the number of errors I found in the articles and books I read. Now being on the other side of the fence, I am now amazed that those authors made such few mistakes in relation to the volume of material they presented.

When covering a topic this broad in scope, there are bound to be mistakes that get by. It is not that extensive efforts are not made to avoid them, it is just that when millions of vehicles are pouring down a production line, each with thousands of possible combinations, there are going to be deviations from the neat categories into which they are supposed to fall.

Also, the work carried on by Ford on the performance projects of the period were conducted on various levels of secrecy. Some were well known, some to this day are still not acknowledged.

In all cases, should the information contained in this book be able to be expanded on through correction of errors or additional facts, it would be welcomed by the author, for only through this exchange of information can we all learn.

PHOTO CREDITS

Consumer Guide: p. 17, p. 23, p. 37 (bottom), p. 38 (left), p. 43, p. 51, p. 52 (both).

Edelbrock Equipment Corporation: p. 135 (top).

Ford Motor Company: p. 24 (upper left), p. 50 (left), p. 53, p. 54 (all), p. 62 (left), p. 67 (top), p. 69 (middle), p. 71 (top and middle), p. 72 (top), p. 87 (middle and bottom), p. 93 (top and bottom), p. 97 (top and bottom), p. 98 (top left and bottom), p. 103 (bottom).

Great Lakes Dragaway-Ray Mars: p. 28 (lower right), p. 31 (upper right), p. 45.

Steve Honnell: p. 129 (both).

International Speedway Corporation: p. 14, p. 38 (right), p. 48 (both), p. 58 (left), p. 61 (right), p. 66, p. 68 (all), p. 74 (top).

Midwest Racing News: p. 133, p. 136 (top and bottom left), p. 145, p. 148 (left).

National Hot Rod Association: p. 18, p. 28 (bottom left), p. 31 (top left), p. 34 (right), p. 61 (left), p. 70, p. 73 (bottom left), p. 136 (bottom right), p. 141, p. 190.

Stuart Sternberg: p. 15, p. 58 (right), p. 112 (bottom), p. 117 (bottom left and right), p. 126 (top), p. 160 (right).

Wisconsin Auto Racing: p. 36 (top), p. 41 (right), p. 46 (right), p. 50 (right), p. 78 (bottom left), p. 104 (bottom), p. 110 (top), p. 113, p. 118 (left), p. 131 (top).

Zecol-Lubaid: p. 39 (both), p. 40, p. 43, p. 56 (both), p. 57, p. 64.

INTRODUCTION

In the years after World War II, Ford Motor Company involvement in the field of performance seldom came in shades of gray. It was either full bore or completely ignore.

This all-or-nothing-at-all philosophy produced some of the finest high-performance passenger cars this country has ever turned out as well as some of the biggest stones that ever clogged the highways.

When the spigot was turned on, Ford not only built and sold cars capable of great things, but raced them as well. In the 1960's the term Total Performance was used in Ford's ad campaigns. It well described the attitude of the company. It seemed there were Fords entered in almost every type of competitive event, from Indianapolis, to the drag strips, to the superspeedways to the sports car courses, with many stops in between.

The booming youth market caught the show and many then marched down to their local Ford dealer and bought a replica of what they just saw win on Sunday.

Lotsa Fords with racing stripes went down the production lines in the performance years. However, some of their owners were surprised, because despite Ford's prowess on the track, there usually was a double standard. Ford made one set of components for race cars and another for the street, where demands were, of course, different. Ford was constantly trying to upgrade its production cars to keep up with the expectations of the buyer—and the competition.

Ford Division performance activity came in three distinct phases after World War II. The first lasted only two years, from mid-1955 through mid-1957 when the Automobile Manufacturers Association ban on performance activity took hold.

The second, which is the subject of this book, started in mid-1959 when Ford found out it was the only one serious about the ban, and ended in 1973 when emissions, safety standards and high insurance rates made performance too expensive for the manufacturers and the customers.

The third phase started about mid-1979 and has grown in scope and stature, right through 1986, when this volume was updated.

This new campaign has exceeded the short-lived first phase and is gaining on the long-running second phase. Ford involvement in racing is again multifaceted and includes NASCAR Winston Cup, several classes of drag racing, Indy Car, off-road, sports car and prototype racing, among others.

The performance enthusiasm, fueled by the Special Vehicle Operations Department (SVO), has resulted in advanced-concept racing machinery, like the Ford Probe IMSA GTP car, and a gradual improvement in the performance level of production cars, like the Mustang GT and SVO, and the Thunderbird Turbo Coupe.

How long the third phase lasts remains to be seen. If Ford follows its own checkered history, it too will be dismantled at some point down the road.

Between the phases there was usually nearly a complete vacuum of performance activity, interesting cars and auto buff enthusiasm. The standard plan was to drop performance-oriented models, high-performance engines and purge parts lists of items not essential to sensible motoring.

This book attempts to sort out the second and by far most interesting phase. Concentration is on the Ford passenger cars of the era and the activity surrounding them. There are also glances at Mercury, Shelby and versions of the Ranchero.

The viewpoint is from the time of introduction for the most part, instead of over the shoulder from the 1980's. The excitement, the disappointment, the race victories and defeats are all presented around the most important theme, the cars themselves. That way the readers can experience them and relate the impressions in terms of their own memories and present environment.

Many of the Fords covered are highly collectable as this is written and no doubt many more will become so in future years.

When a person buys or restores a collector car, the reasons for doing it are as diverse as the variety of cars being collected.

This book can perhaps explain some of the reasons for the popularity of one group of such cars, the Fearsome Fords of 1959 through 1973.

Phil Hall

TABLE OF CONTENTS

CHAPTER ONE
Ford And The Horsepower Race 7

CHAPTER TWO
Thunderbirds That Really Flew 11

CHAPTER THREE
Fast Falcons; Quick Comets 19

CHAPTER FOUR
Total Performing Big Fords and Mercurys 35

CHAPTER FIVE
Intermediates In Action: Fairlane,
Torino And Montego 83

CHAPTER SIX
Phenomena On Wheels: Mustang And Cougar 137

SPECIFICATIONS 176

INDEX 183

Ford Performance—1909 Style

Ford cars were involved in performance activity well before the incorporation of the Ford Motor Company, on June 16, 1903.

On October 10, 1901, Henry Ford and Alexander Winton engaged in an auto race at the Grosse Pointe, Michigan, track. Henry won at 44.8 mph.

In 1902, the famous Barney Oldfield drove Ford's *999* racer. On October 25 the Diamond Trophy races were held at Grosse Pointe. Oldfield got the trophy.

On the cold morning of January 12, 1904, Henry took his Arrow race car onto the ice at Lake St. Clair and established a world's auto speed record of 91.37 miles per hour.

An early example of a team of factory-backed stock car racers is found in 1909 when a transcontinental race was held from New York City to Seattle, Washington.

The first two cars to enter were Model T Fords from the factory. They carried the numbers one and two. Though the challenge was put out to all manufacturers of the day (Ford counted 283 of them), only five cars showed up on June 1 at the starting line. With the Fords were an Acme, a Shawmut and an Itala.

The race was set up by the Alaska-Yukon-Pacific Exposition, the Seattle Automobile Club and M. Robert Guggenheim. First prize was $2,000 in gold and a $3,500 gold cup. Second was $1,500 in gold and that was it.

There was a pace car as far as St. Louis, to keep the racers from taking a short cut. The Fords got there first.

After St. Louis, the 2,700-mile trip to Seattle would be every man for himself. The trip was made interesting by seven straight days of rain. Considering that most of the roads were dirt at the time, one can imagine the conditions encountered. The race became a battle between the Fords and the Shawmut.

Ford number one was driven by Frank Kulick with H. B. Harper as a relief driver. The team was leading by nine hours until American Falls, Idaho, when they hired a guide who proceeded to get them lost in the Great American Desert.

Ford number two, which had engine problems, got back on the road and W. B. Scott and relief driver C. J. Smith took it to victory, breaking the tape at the Alaska-Yukon-Pacific Exposition Grounds at 12:55:35 p.m. June 23.

The Shawmut came in the next day, the number one Ford made it June 25, but had to change an axle after its desert excursion and was disqualified. The Acme arrived six days after the Shawmut and the Itala broke down in Cheyenne, Wyoming, and its owner ordered it shipped to Seattle—in a freight car.

Ford's two entries in the 1909 New York-to-Seattle race waited at the starting line in New York. They were 1909 Model T's, stripped of things like fenders and filled with items necessary for the trip. Behind the wheel of car number two is W. B. Scott and teammate C. J. Smith is next to him.

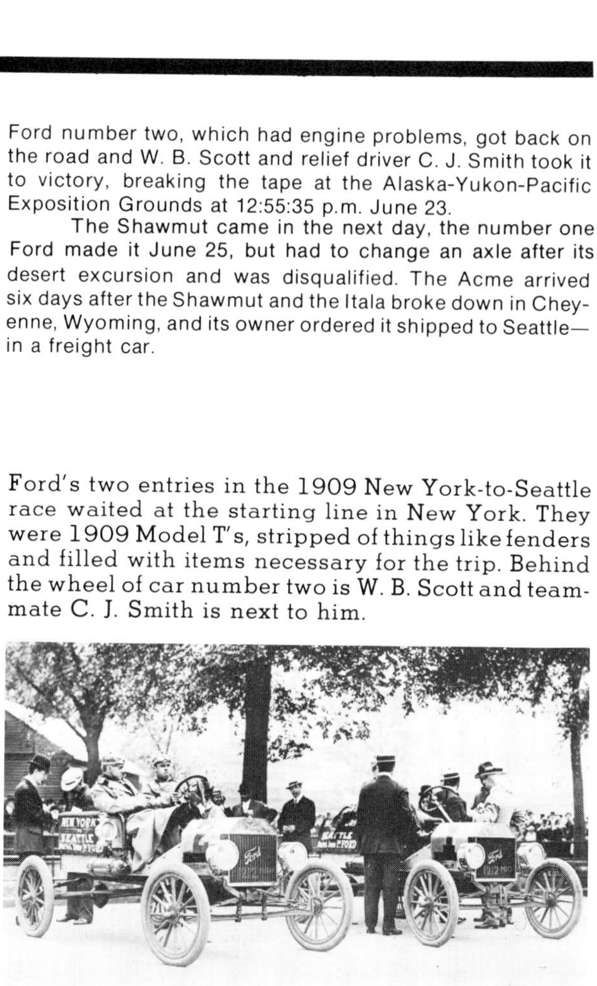

CHAPTER ONE

Ford And The Horsepower Race

Almost from the beginning, Fords seemed to attract a greater than average number of performance and racing buffs. Despite its position in the low-priced field, and for many years a reluctance on the part of the manufacturer to offer speed equipment, Fords were in the forefront of many forms of domestic automotive competition.

When Model T's and Model A's had flathead fours, outsiders designed overhead-valve and even overhead-cam conversions. When the 1932 models came out with Ford's flathead V-8, the wick was turned up even further.

Indeed, many of the early stock car races of the 1930's were composed mostly of modified Fords. Aftermarket speed parts abounded for the V-8.

The sport of stock car racing on oval tracks and on the long straights at the dry lakes in Southern California grew in the late 1930's, early 1940's and again after World War II.

With the hot rodders doing all the work, why should Ford bother to come out with its own performance equipment?

After World War II, the only speed the domestic manufacturers were interested in was that of their production lines, to fill the pent-up demand for new cars. Prewar body and engine designs were put back into production and no one seemed to care. They were more interested in how high their names were on the waiting list rather than how high the horsepower was.

Gradually, new vehicle and engine designs made it to the marketplace. Among the innovations were modern, short-stroke overhead-valve V-8 engines that, through higher compression ratios, could take advantage of the higher octane gasoline that was becoming available. First on the market were the 1949-model Oldsmobile and Cadillac V-8's. Their introduction coincided with the expansion of stock car racing for new cars, which was becoming increasingly popular.

Oldsmobile started what soon became known as the "horsepower race" by putting its 303.7-ci V-8 into its smallest models, resulting in the Oldsmobile Rocket 88. Though having an output of only 135 hp, it was enough to make the 88 the hottest of the 1949 models. They were popular with the stock car drivers right from the start.

The ante was raised with the introduction of the 1951 models when Chrysler brought out its FirePower V-8 with hemispherical combustion chambers and an advertised rating of 180 hp, tops for the industry that year.

Also introduced for 1951 was the Hudson Hornet. Its 308-ci flathead six wasn't exactly a powerplant of the future, but the Hornet had a low center of gravity, was a fine handling car and proved to be the basis for one of the hottest stock cars of the early 1950's. Marshall Teague proved the Hornet could win races and convinced the Hudson factory to support his cars and later other teams' efforts. The idea of direct factory support for late model domestic stock car racing was born!

Make by make, the switch to overhead-valve V-8 engines marched on, with Studebaker getting one for 1951, DeSoto and Lincoln for 1952 and in 1953, Dodge and Buick. Ford and Mercury joined the list with their 1954 models. At 239 ci, Ford's new Y-block V-8 was hardly the thing to burn the tread off your new wide whitewalls. However, for 1954 Ford had the low-priced field's only V-8.

Meanwhile, in the medium- and high-priced fields the battle for bigger engines and higher horsepower ratings went on. Chrysler led the parade with 235 hp in 1954, then set everyone on their door handles with the 1955 Chrysler 300 with that many horses being claimed from its 331.1 cubic inches, which was the same size as the 1951 models.

More overhead-valve V-8's came on stream with the 1955 models with new units for Plymouth, Pontiac, Chevrolet and Packard, the latter also shared with Hudson and Nash.

With Plymouth and Chevy now having V-8 power in the low-priced field, it meant Ford would now have to do some work on its own to maintain its tradition as the performance leader. The hot rodders were no longer enough.

Ford didn't greet its new neighbors lying down. It enlarged the Y-block to two new sizes, 272 and 292 cubic inches, the latter was originally reserved for the new Thunderbird sports car, but was later added to the regular Ford option list.

Of the two new V-8-powered rivals, Chevy turned out to be the biggest threat. Displacing 265 ci, it was lightweight, small and featured stud-mounted stamped-steel rocker arms instead of rocker shafts, which gave the engine a capability to wind up fast, a feature that didn't go unnoticed by the racers, both on the tracks and street.

Coupled with attractive all-new styling for 1955, Chevrolet was an immediate sales success. Adding to that success was something new for Chevrolet, a performance image. After decades of being known as a dependable car with its "stove-bolt six," suddenly buyers, especially young ones, were looking at Chevy in a new light. Chevy was clearly on Ford's performance turf.

While the Chevrolets were no match for the powerful Chrysler 300's, Oldsmobile 88's and Buick Centuries on the bigger stock car tracks, they did do well in some short-track events, where the quick-revving characteristics of the new V-8 were more important than raw horsepower. Chevy scored a couple of minor short-track wins in NASCAR (National Association for Stock Car Racing) competition and the advertising agency played them up like major victories. The public believed.

Ford brass saw what was happening and realized it had better go after the Chevys. With experience at operating a racing organization from the Mexican Road Race days when Lincolns dominated, Ford figured that if it set up a strong effort in NASCAR, Chevy would be silenced.

A pair of factory-backed cars were prepared for the 1955 NASCAR Southern 500 at the pioneer superspeedway, the Darlington (S.C.) International Raceway, a 1⅜-mile high-banked paved track. Crowd interest was high for the Labor Day event and the Fords of Curtis Turner and Joe Weatherly put on quite a show, but front end problems knocked both out and ironically Chevy scored its biggest win ever when Herb Thomas crossed the finish line first in Smokey Yunick's Chevy number ninety-two.

When the 1955 Grand National season was over, both Chevrolet and Ford had two wins. It wouldn't be so balanced for 1956 when the Ford program got rolling with a new 312-ci V-8 as the hottest engine. Fords took fourteen Grand Nationals, second only to Chrysler's twenty-two and far above Chevrolet's three.

NASCAR also had a convertible circuit in 1956 and Fords took twenty-six wins (including nine in a row at one point), to ten each for Chevrolet and Dodge.

During the 1956 season hot options were rushed onto the parts lists by Ford, Chevrolet and others, as the hottest combination was sought in the most fervid part of the horsepower race.

Fords of the 1930's laid important groundwork for Ford's performance image in the low-priced field. Examples like this 1939 coupe were hopped-up and raced before and after World War II.

The first production Fords with overhead-valve V-8 power were the 1954 models, which got a 239-ci Y-block, and rating of 130 hp. The size only lasted one year, as 1954 was the last time Ford had the V-8 low-priced field to itself.

The 1957 models reached new highs for both Ford and Chevrolet. Chevy enlarged its V-8 to 283 cubes with the top option being mechanical fuel injection, which had a rating of 283 hp. Ford countered with a McCulloch supercharger for its 312 V-8, with 300 horses advertised.

Ford took cars to the Bonneville (Utah) Salt Flats and broke 458 national and international performance records in September of 1956 and timed the announcement to coincide with its all-new 1957 models.

Many engines were now rated well past the 300 mark, with an option for Chrysler's 300C topping the pack at 390 hp.

Car enthusiasts may have been in their glory, but there were some dark clouds on the horizon. Getting as much publicity as the horsepower race was the highway death toll. Do-gooders in Washington were talking about cracking down on the auto industry, maybe even regulating the way cars were made.

Also, the cost of fielding racing teams and the ever-changing technology's thirst for new parts and development was an increasing burden on those manufacturers seriously playing the high-performance game.

In early 1957, the Automobile Manufacturers Association (AMA) began looking at ways to gracefully back out of the horsepower race. Sensing trouble ahead and wanting to slow things down a bit, NASCAR banned all but single, four-barrel carbs in the Grand National in April of 1957. This wiped out Chevy's injection and Ford's supercharger.

It was only part of the slowdown, as the AMA ban, announced in early June of that year, put an end to factory-backed racing teams and the promotion of high-performance models.

Teams were disbanded, with parts being sold or given to the various parties under contract.

Ford sold much of its equipment to John Holman and Ralph Moody of Charlotte, North Carolina. To help earn money to set up shop, Moody raced one of the supercharged Fords in United States Auto Club (USAC) stock car competition in 1957 (where the setup was legal).

Despite the pull-out, Fords did very well in NASCAR, winning twenty-seven Grand Nationals to Chevrolet's eighteen.

The immediate effect of the AMA ban was the factory pullout from stock car racing. The 1958 models were too far along for major changes, with several new engines being introduced as a natural progression of the horsepower race.

Hotter versions of the engines, which usually were introduced after the first couple of months of the model run, were omitted from many of the 1958 makes, however.

Ford's new powerplant for the 1958 models was its "FE" big-block, coming in 332- and 352-cube versions. The Y-block, at 312 inches, had been stretched about as far as it could go, and increasing cubic inches was the cheapest way of gaining horsepower and performance.

Top rating for the 352, which would be available in both regular Ford and Thunderbird lines, was 300 hp. Compression ratio was 10.2:1 and it had solid lifters. But Ford was not alone with new bigger engines; Chevrolet brought out its 348-cube V-8 big-block and Plymouth got a 350-cid version of the new Chrysler B-block.

Styling for the 1955 Fords featured a major face-lift and a new roofline for the Fairlane Crown Victoria shown here. Size of the V-8 engines that year started at 272 and rose to 292 cubic inches, in reaction to competition from Chevrolet and Plymouth.

Using the basic roofline of the Crown Victoria, Ford's 1956 Fairlane Victoria was an attractive car, especially with the two-tone paint and skirts, as shown. Maximum engine size remained 312 ci, with a top rating of 225 hp.

Each manufacturer reacted differently to the AMA ban. Pontiac slowed its pace the least, continuing to develop and sell performance parts and indirectly support racing teams.

Chevrolet went underground, Chrysler altered its programs, using law enforcement car development as a guise.

Then there was Ford. Except for American Motors and Studebaker-Packard, which were inactive for most of the horsepower race anyway, Ford took the AMA ban the most seriously.

It dropped performance parts from its lists and cut the options on its new cars. There was no hot 352 introduced mid-year and even the performance axle ratios were chopped. After the start of the model year, the 352 compression ratio was cut to 9.6:1 and hydraulic lifters replaced the solid units.

In contrast, Chevrolet added a 315-hp version of the 348 in mid-year, thirty-five ponies above the original rating, thanks to three two-barrel carburetors. The engine was supposedly for law enforcement purposes.

Pontiac added a 330-hp Tri-Power job, well above the 310-hp fuel-injection setup, which started out as its top output engine.

Motor Trend tested a 1958 Fairlane 500 two-door hardtop with the 300-hp 352 and got a 0-60-mph time of 10.2 seconds. In contrast a 280-hp Chevrolet made the trip in 9.1 seconds and a 305-hp Plymouth in 7.7.

Word of mouth and in print spread quickly of Ford's plight. Slow days were ahead for Dearborn performance fans.

Highlighting Ford's first performance period was a supercharger option for the 1957 312, which produced 300 hp. Fireball Roberts is shown in the 1957 Daytona NASCAR 160-mile Grand National in a supercharged Custom two-door sedan. NASCAR banned the setup a few months after the February event.

CHAPTER TWO

Thunderbirds That Really Flew

The only true high-performance Ford Thunderbirds to date were the two-place 1955 through 1957 models. In each of those years, the small Birds were first in line for the most potent engines Ford Division dished out in the hectic early days of the so-called "horsepower race." However, Turbo Coupes of the mid-1980's are coming close to the early Birds in the go-fast category.

Culminating the early Thunderbird powerplant story was the 1957 F-Bird with a McCulloch supercharger extracting 300 advertised horsepower from its 312-ci Y-block V-8. Production of the 102-inch-wheelbase Thunderbirds ceased a few months after the Automobile Manufacturers Association (AMA) June 1957 ban on the racing and promotion of performance cars by the manufacturers.

Replacing the early birds were the all-new four-place 1958 models, which were about as different from their predecessors as you can get and still end up in the same general market area, the personal sporty/luxury car.

Comfort and luxury were far higher priorities than all-out performance or sportiness for the new four-seat models. Yet, before their three-year styling cycle would end, the "Squarebirds," as they later became known, would play an important role in the return to the business of building, selling and racing high-performance cars by Ford.

The new 1958 Thunderbird was larger in every way over the two-seater jobs. Wheelbase grew from 102 to 113 inches, length from 181.4 to 205.4, width from 72.8 to seventy-seven and weight ballooned more than 700 pounds to around the 3,900 mark.

The increases were not without reason, as seating capacity doubled, with four stylish bucket-type seats replacing the old model's single bench. The result was a car unique in its class, without direct competition.

Power came from the newly introduced 352 V-8, rated at 300 horsepower. There were no options.

Chassis design changed just as drastically as concept. The old Bird's frame was replaced with unitized construction, which it shared with the all-new 1958 Lincolns, which were built alongside the T-Birds at a new factory in Wixom, Michigan.

Instead of leafs, the rear suspension was now coils for a softer ride.

A two-door hardtop was first introduced, followed by a convertible. In the overstyled period into which it was born, the new Thunderbird had rather clean lines.

Performance was nothing to brag about, as the new Bird was about 500 pounds heavier than a comparable Ford. *Motor Trend* pulled 0-60 in 10.1 seconds.

However, the T-Bird was a trendy car and sold well in the depressed market of 1958. Despite a late introduction, 37,892 Thunderbirds were made for 1958, compared to the 1957 run of 21,380.

While the four-passenger Thunderbird contributed badly needed money to Ford's bankroll, it did nothing for the division's sagging performance image.

When the conservatively styled 1959 Ford passenger cars were introduced in the fall of 1958, the power story was even sadder. Compression ratios and horsepower ratings were cut on most engines. The top 352 was still rated at 300 advertised horses, but its compression remained at 9.6:1.

Motor Trend was still standing on the gas and now a 1959 Fairlane 500 dropped to 10.5 seconds for a 0-60 jaunt. A Chevrolet Impala made the trip in nine seconds flat and a Plymouth Fury was another half-second faster.

While the value of a magazine's time report is limited at best, it indicated Ford's relative slip in performance stature in the late 1950's.

When the 1959 models were introduced, there was no performance program within the confines of Ford Division. The AMA ban was being adhered to, to the letter, although there was suspicion that the competitors were increasingly violating the agreement.

Without planning it that way, Ford did take its first step on the road to recovery with the 1959 Thunderbird. While a new grille, side trim piece and rear taillight panel were the main visual differences, some important changes were made under the skin.

In a rather simple move, Thunderbird had an option of the largest engine in the passenger car field, 430 ci. It was rated at 350 hp; on paper, higher than all but a few of the high-priced luxury cars and much higher than anything in its class. At 430 cubes, the new Thunderbird Special 430 V-8 was bigger than anything Ford Division would sell for the next dozen model years.

Actually, at least one prototype 1958 Thunderbird was built with the 430, which was then rated at 375 hp. Before the launch was delayed for the 1959 models, it was featured in revised (March 1958) Thunderbird deluxe sales catalogs.

While the numbers look nice, the real life facts were considerably less spectacular. There was an excess of Lincoln engines and since Lincolns and T-Birds were all made in the same plant, Ford offered the Lincoln 430-ci powerplant as an option.

The 350 horses peaked out at a lazy 4400 rpm. Torque was rated at 490 pounds-feet at 2800 rpm. A 10:1 compression squeeze was at least higher than the regular Fords.

It came bolted to an automatic called Cruise-O-Matic, using the name of the three-speed automatic introduced with the 1958 models, but the transmission was actually the heavier unit installed in big-engine Mercurys.

The 430 listed for $177 extra, but you also had to pop for the automatic at an additional $230.80.

Thunderbird literature that year wasn't especially shy about describing what a 430 could do if you opted for one: "Thunderbird's new optional 350-horsepower engine gives you the ultimate in action—a luxury car that is the peer of even sports cars in performance!"

Still standard was the 352 Special V-8 at 300 hp. It came with a three-speed manual transmission—overdrive and Cruise-O-Matic were optional.

Another new feature of the 1959 T-Bird was the rear suspension. Actually it was an old feature, as leaf springs returned. The coils just did not work out for handling and the leafs were cheaper, gave more control, took the output of the 430 better and didn't sacrifice all that much in ride.

While the majority of the car-buying public could care less about the 430 or leaf springs, car buffs took notice of the 1959 Thunderbird.

Hot Rod got hold of a 430 Bird and despite weighing in at 4,200 pounds, it did 0-60 in nine seconds, 1.5 better than a 352-equipped version. The big-engine job turned the quarter-mile from a start in seventeen seconds and hit the traps at the end at 86.57 mph. Not enough to warrant going out and buying a trophy case, but

Single-seat, first-generation Thunderbirds had engines available up to the 300-hp supercharged 312. While great for Ford's image, they did not generate the sales figures needed to keep them in production. Shown is a 1957 model, the last before the four-seaters took over.

"Squarebirds" replaced the single-seaters starting with the 1958 model year. A convertible is shown going through the stages of putting up the top, which was stored in the trunk. The 1958 models only came with one engine and had coil rear springs.

enough to blow away most of the showroom-stock 1959 Fords. The standard 2.91 gears also added a bit of respect for the performance numbers.

The magazine noted that in standard form, the 430 Thunderbird was not a screamer, but that a change of axle ratios would improve the times and with proper tuning, "there is no reason why the Bird shouldn't be capable of holding its own with the other similarly equipped stockers running at the weekend drag races."

Lincoln engines weren't noted for spawning shelves full of hop-up equipment, but the 430 was the basis for an earlier 1958 Mercury option rated at 400 hp. That version never reached production stage, but the triple two-barrel intake manifold that was supposed to be installed, was sold by Bill Stroppe of Long Beach, California, who was the factory racing representative for Mercury before the AMA edict.

While the 1959 Thunderbird never did establish itself as a contender in drag racing, it was a success in stock car racing on oval tracks.

Very aware of the potential of the 430 in a small package like the 113-inch-wheelbase Thunderbird, Holman & Moody went into the business of building and selling Thunderbird stock cars.

NASCAR didn't like the idea of such a big engine in a car smaller than the competition, but did want to keep the Ford fans happy and so let the 1959 T-Bird run in its Grand National and Convertible stock car divisions.

Holman & Moody was an independent business at the time, receiving no help from Ford, but through contacts at the plant, they were able to purchase several scrapped Thunderbirds that didn't make it to the end of the production line. The price was right and they hauled the bodies back to Charlotte and built up stock cars.

The cars used the reinforced unitized construction with a roll cage installed. Some came with "zipper tops" which meant the roof could be unbolted for the convertible races and left on for the Grand Nationals.

The Lincoln engines were used and instead of the automatic, a heavy-duty three-speed manual transmission was used, a move that was legal.

Focal point of each new NASCAR racing season was traditionally the February festivities on the sand at Daytona Beach, Florida. Speed trials were run on the hard beach and an oval was created utilizing the sand for one straight and an old narrow road for the other, to form a 4.1-mile course for stock car races.

This setup was used each year for Speedweeks activities from 1949 through 1958. However, it would be a different ballgame in 1959, as only the Speed Trials would be held on the sand.

Nearby sat the showplace of stock car racing, the 2.5-mile paved, high-banked Daytona International Speedway which was now ready to hold the closed track stock events.

Darlington pioneered the superspeedway idea for stock cars in 1950 and with the coming years more of the giant speed plants would be built in the Southeast to bring NASCAR racing up to the status of a major league sport.

The exposure of stock car racing to enormous numbers of fans was something the automobile manufacturers would have a hard time ignoring in future years.

Interiors of 1959 Thunderbirds sported an unmistakable feel of grand, custom-built luxury, with many functional power assists and accessories.

Mild restyling marked the outside of the 1959 Thunderbird, like a straight-bar grille, new side trim and new wheel covers, but more important were the changes underneath which included a new 430-cid, 350-hp V-8 and a new rear suspension.

The new Daytona track was ready for its share of Speedweeks activity in February and so was Holman & Moody, having constructed eight 1959 Thunderbirds in time for the festivities. Six of them were sold to other racers for $5,500 complete.

The first test for the new cars was qualifying for the 500. Cotton Owens, the 1957 Daytona Grand National winner, set fast time in a 1958 Pontiac at 143.198 mph. Former Ford factory driver Curtis Turner was next in a T-Bird at 142.993, followed by Lee Petty in a 1959 Oldsmobile at 141.709 and Chicago driver Tom Pistone in another Thunderbird at 141.376.

In all, Thunderbirds took three of the top five qualifying spots. A hundred-mile qualifying race was held and three-time NASCAR convertible champ Bob Welborn of Greensboro, North Carolina, won it in a 1959 Chevrolet; but Thunderbird drivers took three of the next four spots with Fritz Wilson of Colorado second, Pistone third and Peruvian driver Eduardo Dibos fifth. Fourth-place Joe Weatherly, also a former Ford factory pilot, broke up the trio in a Chevrolet.

The pre-500 events indicated that despite the lack of any factory help, the Thunderbirds would be highly competitive and put Ford in the spotlight.

Pistone reflected on the Daytona events in 1959. "The Thunderbird was heavier, we had to make up for it with cubic inches. The Pontiacs were faster, but we could keep up with the Chevys. They could run 6400 to 6500 rpm, while the Lincoln engine would blow over 6000."

The 430-ci Bird compared with 389-inch 1959 Pontiacs, 348-inch 1958 and 1959 Chevrolets and Petty's 394 inches of Oldsmobile.

With a crowd estimated at 47,000 in the stands, the fifty-nine-car field tock the green flag. Welborn started on the pole in his Chevy and led the first lap, Pistone took over for two, Welborn grabbed it back for four, then it was Pistone again and so it went.

The fans loved it with Fords, Chevrolets, Pontiacs and other makes mixing it up mile after mile.

All went fine until the half-way mark when tire-chunking problems struck, with all makes affected. The tire companies had no real way to prepare for the event and the worst was realized.

"The drivers that went the fastest had the most problems and since we ran too hard, we had problems," Pistone said.

Pistone, Turner and Tim Flock all had their T-Birds slowed with rubber problems. Turner had to stop a dozen times just to change tires.

Among those not affected by tire troubles were Lee Petty, the 1954 and 1958 Grand National titlist, and Iowa Thunderbird jockey Johnny Beauchamp—they literally battled down to the wire. The winner was decided three days later after a photographer's shot of the finish was developed, with the nod going to Petty.

Other T-Bird pilots didn't fare as well with Pistone eighth, Flock ninth and the beleaguered Turner an unlucky thirteenth.

Nevertheless, the point had been made, Ford wasn't quite the stone the detractors and competition felt it was.

Not only were 47,000 fans at Daytona, but so were a number of factory personnel from the manufacturers, especially Chevrolet and Pontiac.

When Ford found out it was the only one in the AMA edict boat, it began anew a performance program in early spring of 1959. And one thought must have come to Ford's mind: Chevy and Pontiac were helping the racers by offering the right speed equipment, yet the totally independent Thunderbirds could keep up and almost win—just think what could be done if we get back into the business!

In the course of the 1959 season, conducted mainly on shorter tracks where the lighter cars had the advantage, Thunderbirds took six wins. Chevrolet, which fared well on the short tracks, led all comers with fourteen wins, followed by Plymouth (which picked up Lee Petty and his son Richard mid-season) with nine, followed by regular Fords (1957 and 1958 models) at eight, then T-Bird.

Returning to Daytona July 4 for the Firecracker 250, the Grand National gang saw just how competitive the T-Birds were on a bigger track. Fireball Roberts scored Pontiac's only Grand National win of the year, but six of the top nine cars were Thunderbirds.

For some reason, NASCAR never classified the 1959 T-Bird wins as Ford wins, but kept them separate, even though Thunderbirds won events in later years and the victories were added to the Ford tally.

The Thunderbirds were a one-year phenomenon in NASCAR. When the 1960 Ford Starliner and its 360-hp 352 was introduced for the 1960 season, the Squarebirds were quickly dropped by the front line teams. Holman & Moody switched to building Starliners.

Lesser teams continued to race the 1959 Birds in the 1960 and later seasons, updating some to 1960 trim, but they were no match for the powerful new Fords and other makes in stock car racing.

1960

Thunderbird for 1960 was much the same as the 1959 models, being on the third and final year of its styling cycle. The addition of an extra pair of taillights and a complicated grille pattern visually were the most noticeable differences with their main function to make people appreciate the 1959 models.

A new steel sunroof was an option at $212.40 and gave T-Bird's luxury-oriented buyers something more to add to hardtop models.

Back on the option list was the 430 but, as in stock car racing, it no longer held the fascination of the Ford performance fan.

Motor Trend tested a 430-equipped 1960 Thunderbird and got an 8.9-second clocking to sixty. Meanwhile *Motor Life* reported that a 360-horse Starliner did the same thing at 7.1 seconds and added that a top speed of 152.2 was reached.

At this point, it would be fair to say Thunderbird had served its interim role well and move on, but Thunderbird fans would never forgive if the Sports Roadster wasn't acknowledged, even if it was no longer at the forefront of Ford's performance armada.

For some reason, Detroit (and Dearborn) liked to do styling cycles in three-year periods in the 1950's and early 1960's. The practice dictated a three-year life for many styles, be they a beauty or a turkey.

1961

While the 1958-60 Thunderbirds probably fell somewhere between the above criteria, they were replaced by the all-new 1961 models.

Styling was smoother and the grille/bumper bore more than a passing resemblance to factory-mate Lincoln Continental. Width was down a little over an inch, length reduced a fraction of an inch, and treads increased with the front an inch wider and the rear three, now at sixty-one and sixty, respectively.

Weight went up about 150 pounds, but the power to haul it around went down. Only one engine was available, Ford's new 390-ci version of the FE engine, achieved by a bore and stroke increase on the 352.

Even though it gained thirty-eight cubes, Ford retained its 300 advertised horsepower rating (which still was quite optimistic), but did increase its advertised torque from 395 pounds-feet at 2800 rpm to 427 at the same speed. Gone were the standard transmissions; Cruise-O-Matic was now standard.

Production of the 1960 Thunderbird set a record at 92,843, of which the 1961 model at 73,051 fell far short. However, a soft new-car market during the latter year was probably more to blame than the new styling.

Nevertheless, the market was rapidly shifting back to performance and sporty cars. And Ford dealers never forgot the excitement the early 1955-57 Thunderbirds created. While the volume of such a design could not justify the tooling costs, Ford did make a move to recapture some of the magic of bygone days.

1962

When the 1962 Thunderbirds were introduced, there were four models instead of just the hardtop and convertible. At the luxury level was the new Landau hardtop and the sporty model was the Sports Roadster.

To achieve a two-seat arrangement, a fiberglass tonneau cover for the rear seats, complete with headrests for the front seats, was included. It was removable and could be kept in place with the top up or down.

After they were no longer front-line equipment in NASCAR, the 1959 Thunderbirds saw service with lesser teams and on lesser circuits, usually updated to 1960 models. This example is shown being driven in a 1961 USAC event at the Wisconsin State Fair Park Speedway by Rich Sutkus. Note that although the grille is from a 1960 model, the side trim on the spear is still of 1959 origin.

To add to the sporty effect, it also included Kelsey-Hayes wire wheels, in chrome with knock-off-type hubcaps. Rear fender skirts were omitted to show off the wheels. Special medallions under the Thunderbird name on the front fenders and a passenger grab bar on the dashboard further distinguished the Sports Roadster.

With the low lines of the Thunderbird, the bootless rear deck and the rear cover in place, the Sports Roadster was indeed a fine-looking machine and did recapture some of the flair of the first Birds.

Even though it was available in the line from the start of the model year, the Sports Roadster was promoted heavily in Ford's Lively Ones campaign. It did have something new to offer at the time, an optional engine setup.

Added to the option list part way through the 1962 model run was the Sports V-8, a 340-hp version of the 390. The increased numbers were made possible by a new intake manifold similar to that on the 401-hp 390 in regular Fords with three two-barrel carbs.

Compression ratio of the Sports V-8 went from 9.6 to 10.5:1 with a rating of 340 hp at 5000 rpm. Torque stayed at 427 pounds-feet, but was now achieved at 3600 rpm, up 800. The Thunderbird performance engine retained the hydraulic lifters of the 300-horse 390, while the hot regular Ford engines utilized mechanical units.

Sales of the Sports Roadster, which had a base price in 1962 of $5,439 ($651 over the convertible), weren't nearly as spectacular as the car, as only 1,427 were made as 1962 models. However as a sporty image car, Ford had a winner all the way.

1963

A minor facelift marked the 1963 models, with the first noticeable sheetmetal change for the styling cycle. A ridge extended over the front wheel opening from the tip of the front fender to about three-quarters of the way through the door. The device added a half inch to the overall width. New grilles and other trim also marked the 1963 Thunderbird entry.

The Sports Roadster was back again, but wasn't heavily promoted. Incidentally, those Kelsey-Hayes wire wheels were an option on other T-Bird models at $373.30 a set, an expensive option in 1963, but only a fraction of what they would sell for on the collector market in later years. Also returning was the Sports V-8, but it was quietly dropped from the option list mid-year as T-Bird was preparing for its luxury car career deserting its role as a sports-type car once again.

Only 455 Sports Roadsters were called for in 1963's model run, compared to 14,139 Landaus.

The regular Thunderbird hardtop for 1962 was altered little from the 1961 model with side pieces and grille pattern changing; the big news was the addition of two new models. Trying to recapture some of the magic of the two-passenger 1955-57 Thunderbird, Ford brought out the Sports Roadster. The removable fiberglass tonneau cover had built-in headrests. When combined with the disappearing top, it did make a good-looking package. Helping in the looks department were Kelsey-Hayes wire wheels with knock-off-style hubcaps.

1964

On schedule, a new body shell was introduced for the 1964 Thunderbird. It featured more angular sheetmetal covering a package that was a fraction of an inch larger in wheelbase, width and length over the 1961-63 models.

Interior styling was changed with wraparound rear seats and an aircraft-like cockpit for the driver. Luxury and gadgets were stressed over performance, which was further reduced by added weight and only the 300-advertised-horse 390 to pull it around.

Motor Trend tested one with a curb weight of 4,740 pounds. The testers got it up to sixty in 11.2 seconds and said it was good for a top speed of 105 mph.

Gone from the model lineup was the Sports Roadster, but some of the pieces were still around. The fiberglass tonneau cover and Kelsey-Hayes wires were available, individually, as options.

Only about forty-five cars left the factory with those options, making the 1964 Sports Roadster-optioned version the rarest of the three years.

Aftermarket shops have been producing the tonneau covers and even some of the trim pieces, so not all of the Sports Roadsters around today are originals.

The major contribution of the Thunderbird to Ford performance continues to be the Lincoln-engined 1959 models, when Ford had little else to offer and needed something to fill the gap before its performance program could get moving again.

The second and final year for the Sports Roadster as a model was in 1963. Basically the same as 1962, visible changes include the flared streak down part of the side and door trim, plus a new grille pattern. This early photo shows a plain front fender, without the badge later models got.

With performance a low priority, Thunderbird stylists and engineers concentrated on fancy interiors and gadgets for the 1964 models. The interior of the convertible is shown here, note the wraparound rear seats.

New styling marked the 1964 Thunderbird with more squared-off lines. Sports Roadster pieces were available as options, but the model was discontinued. Only about 45 convertibles were so equipped.

Old two-passenger Thunderbirds remained active in drag racing for many years after production ceased with the 1957 models. Here Bill Coons pulls off the line at the 1966 NHRA Nationals in his sohc-powered 1957 Bird.

1977-1982

After reaching gigantic proportions in the 1967-76 span, Thunderbird was downsized for the 1977 model year, sharing chassis and much bodywork with the intermediate LTD II. Starting with the 1980 models, Thunderbirds got even smaller, using a stretched Fairmont chassis and slab-sided bodywork. While neither of these models were to be considered performance cars, the body styles were used by NASCAR Grand National and other stock car racing teams, marking the return of the Thunderbird to the sport of stock car racing, where it was last a factor in 1959.

1983-1987

Thunderbird got a new aerodynamic coupe body for model year 1983, and the sleek design was a center point in Ford's plan to be competitive in NASCAR. The style also adapted well to the popular Pro-Stock class in drag racing.

At first the powerplant lineup for the 1983 Thunderbird was plain vanilla, but in April of 1983, a performance version, the Turbo Coupe, was added, with a turbocharged, fuel-injected version of Ford's 140-ci-four standard.

Considering the still large size of the Thunderbird and its aim at the luxury-oriented buyer, there were questions as to why a high-output V-8 was not part of the package. However, the Turbo Coupe is still with us as the 1987 model year starts, and a hot V-8 Thunderbird has yet to reach production status.

Styling remained constant with only detail changes from 1983 through the 1986 model year. For 1987, the Turbo Coupe got a sloped nose, while other models got a revised, more aerodynamic grille. Rear quarter and rear deck styling was also revised for 1987.

The NASCAR Ford teams gradually made the Thunderbird aerodynamics work, with the pinnacle coming in 1985. When Bill Elliott won the September 1, 1985, Southern 500, he received an unheard-of $1 million bonus from the R. J. Reynolds Tobacco Company for taking three races in its Winston Million program.

Not long after the introduction of the all-new 1983 Thunderbird in the fall of 1982, a NASCAR Grand National version was shown, making the car's aerodynamic intent known. However, it was Bill Elliott who made the Thunderbird famous in 1985 when he won a record 11 superspeedway wins in NASCAR Grand National competition. He topped $2 million in winnings that year.

Thunderbird got rid of its boxy lines with this sleek 1983 model, which would eventually return it to the performance field.

The Thunderbird performance car for the street from mid-1983 through 1987 was the Turbo Coupe, shown here in 1984 form. Four-cylinder power was used all years.

Showing off its new flush headlights, sleek nose and other changes for 1987 is the Thunderbird Turbo Coupe. The styling tricks were aimed at regaining supremacy in NASCAR superspeedway racing.

CHAPTER THREE

Fast Falcons; Quick Comets

No one can accuse Ford of having performance in mind when it designed its entry in the compact car field in 1960, the Falcon. Yet after a slow start, the simple, basic design was utilized to establish Ford's presence in the performance compact market and hold a place until the sporty Mustang came along.

In the late 1950's, major automotive changes were taking place in the domestic marketplace. Buyers were rejecting big cars and purchasing small economy cars in increasing numbers. American Motors was right on target with its revived 1958 Rambler American, Studebaker's 1959 Lark sold well and the stage was set for the introduction of the Big Three compacts for the 1960 model year.

In the fall of 1959 the first compacts from Ford, Chrysler Corporation and General Motors bowed, the Falcon, Valiant and Corvair, respectively.

Chevrolet's Corvair was the most technically different with its rear-mounted air-cooled engine, fully independent suspension and rear transaxle. Valiant (which was sold by Plymouth dealers, but would not carry the Plymouth nameplate until the 1961 models) was a little more conventional with a front engine, but had it slanted thirty degrees. Valiant's styling was the most radical of the three.

The conservative member of the group was Falcon. Both styling and engineering were simple. The engine was a front-mounted lightweight, short-stroke six, displacing 144 cubic inches and advertised at ninety horsepower. Front suspension was via coils and shocks, mounted over the upper A-frame. Leaf springs did the job in the rear. Three-speed manual or two-speed Fordomatic transmissions were mounted in the usual place, behind the engine.

Body construction was unitized and the sedan package came with a 109.5-inch wheelbase, overall length of 181.2 inches, width of 70.1 inches and height of 54.5 inches. Front tread came to 55 inches, with a half-inch less at the rear.

Lines were tasteful and simple with a concaved grille, shallow concaved side section running the full length and ending at the rear with Ford's familiar bull's-eye taillights.

1960

Initially a two- and four-door sedan were produced in standard and deluxe trim, but in early 1960 two- and four-door station wagons were added, as was a pickup truck version, the Ranchero. Previous Rancheros were in the full-sized category. Wagons were longer than the sedans with 189-inch overall lengths.

Corvair took the spotlight when the cars came to market, but it was Falcon that became the hottest seller. Falcon even topped Rambler to become the most popular compact for the 1960 model year.

While Falcon may have been the best performer in the sales race, it was nothing to brag about on the street. Even a three-speed version couldn't break twenty seconds in *Motor Trend*'s 0-60 trials. Valiants were in the seventeens and Corvairs were about on par with Falcon.

Falcon was conceived in the economy-conscious days of 1958 and 1959, but born in the 1960 model year, when the interest in performance was enjoying a rapid comeback.

While the average Falcon customer wasn't overly concerned that the little six really only put out about 65 hp, the racers and hot rodders were busy looking for some punch for the newly emerging sport of compact stock car racing.

Ford didn't exactly make it easy for them, as the Falcon's intake manifold for the single-throat Holley carburetor was cast into the head, a practice Rambler had followed for years.

To get the little six to put out more horses, more fuel and air had to get in. Three different companies outside of Ford approached the problem and all three, Holman & Moody, Edelbrock Equipment Company and Bill Stroppe & Associates, all came up with the same basic solution: three Falcon carbs on a long intake manifold.

The top of the stock intake runner was cut open for installation of the two extra carbs. Holman & Moody cut holes for the end carbs, but the other two setups involved removing the top and adding a new cast piece to mount the units. Holman & Moody's setup was mainly for racing and all three carbs worked at once, while Edelbrock and Stroppe had progressive setups with the end carbs opening later than the center one, a more streetable arrangement.

Valve springs were replaced with stronger units, compression ratios were raised by shaving the heads and exhaust manifold changes, including headers, were among the other improvements made to help the 144's output.

As the model year passed, more exotic modifications included Hilborn fuel injection and Paxton superchargers.

Stroppe, who was under contract to do special vehicle work for Ford, sent his Falcon triple-carb setup to Dearborn for evaluation. For a short time the setup, which produced 128 hp at 6000 rpm, was considered for a regular production option, but the idea was dropped, probably because of compact car racing results and, like the others, the Stroppe setup was available as a parts kit.

Great things were envisioned for the new domestic compacts in the world of auto racing, yet their promise was not fulfilled due to a number of factors.

The cars were not powerful enough compared to the bigger stock cars to put on a decent show on the oval tracks, especially the superspeedways. And they were no match in the handling department to compete against the sports cars and imported sports sedans on road courses. The early compact car events, all held on road courses, did nothing to help the cause.

In early November a six-hour compact car event was hastily put together at the Continental Divide Raceways near Denver, Colorado. Due to the lack of domestic entries, foreign small cars were added. No Larks or Valiants were entered.

With no mid-year performance packages available, the Falcons were in stock 90-hp trim and the Corvairs, with 140-cube engines, had ten fewer.

When it was over, a Rambler American, co-driven by Johnny Mauro and Tommy Rice, took the win, followed by a 36-hp Volkswagen Beetle, two Falcons and a Volvo. The Corvairs had tire problems and placed sixth and seventh.

A December 12, 1959, two-hour contest at Sebring, Florida, over the famed 5.2-mile airport course had better representation from the domestic compacts, but liberalized rules ruined any chance for fair competition. Jaguar 3.4 sedans and Lark V-8's ran with predictable results. Walt Hansgen won in a Jaguar, followed

Starting with a simple, basic approach in both styling and engineering, Ford Falcon made its debut as a 1960 model. At first only two-door (shown here) and four-door sedans were offered. Falcons were entered in early compact car races with little success.

A second Ford Motor Company entry in the compact car field came mid-year in 1960 when the Comet was introduced, using a stretched Falcon chassis and shared mechanical components. Early Comets were underpowered, but later models would make up for that.

by Curtis Turner in a Lark, Ed Crawford in a Jaguar and Fireball Roberts in a Lark. A Volvo was again fifth and a Corvair sixth. Denise McCluggage brought the field's lone Falcon in eighth.

Not exactly of the off-the-street variety, the Jags were prepared and entered by Briggs Cunningham and the Larks by none other than Holman & Moody.

Perhaps the crushing blow to the hopes of the Falcon and Corvair in compact car racing came in the next event, at Daytona, when Valiant's Hyper-Pak option blew the competition away in convincing fashion.

Falcons had the tri-carbed setups, Corvairs the newly announced 95-horse option and the Lark V-8's weren't allowed. However, no one quite expected the Valiant bombs.

In stock form, Valiant had inches on the Corvair and Falcon with a 170-cube slant-six standard, advertised at 101 hp. However, the optional (over the parts counter) Hyper-Pak brought the advertised horses up to 148, and reports had the racing engines into the 180's. Since the engine was tilted thirty degrees to the right, it left room for long intake manifold runners. Using Chrysler's newly gained experience with ram induction, the tubes were tuned, equipped with a four-barrel Carter carb and complemented with a special cam, high-compression heads, stiffer valve springs and dual-point ignition.

Valiant chassis were beefed-up, brakes and tires were at racing strength and top drivers were hired.

Two events were held at Daytona International Speedway on January 31, 1960. The first, televised nationally by CBS, was a ten-lap contest over the 3.81-mile combined oval and road course. The second went for twenty laps on the 2.5-mile high-banked oval.

Marvin Panch led seven Valiants across the finish line for a rout in the first event; Joe Weatherly, in eighth, was the best Falcon effort; followed by Ricardo Rodriguez, the famed Mexican sports car driver, in a Corvair.

The oval track portion wasn't much different, with Panch leading a one-two-three parade for Valiant, averaging 122.282 mph. Weatherly came in fourth, but was a lap behind the high-flying Valiants. Turner took fifth and short-tracker Jim Reed came in sixth in a Corvair.

With the whole country knowing about the compact car mismatch, the future of compact car racing dimmed considerably. It was after the Daytona disaster that Ford decided not to offer the Stroppe setup as an RPO option, and Falcon performance development looked pretty dead.

The Falcon six did see some racing applications other than stock car competition. The engine was adapted to midget auto racing, with fuel injection being used. Ironically it was used to replace another Ford powerplant, the prewar V-8 60, that was popular in midget competition.

A Falcon six was also put in a Streamliner. Bored to 156 cubes and equipped with Hilborn fuel injection, an Isky cam and magneto, the powerplant was good enough to give *Sports Car Illustrated*'s Bill Burke a new Class D record of 205.949 mph at the Bonneville Salt Flats, up 43 mph over the old standard for the class.

Falcon was given a big brother before the model year was out, as the Comet from Mercury made its debut in the spring. (It would not be the Mercury Comet until the 1962 models came out.)

The first Comets to carry the Mercury nameplate were the 1962's. Like Falcon, Comet had its sports model with bucket-type seats and console similar to the Futura, the S-22. Comets also shared Falcon's six-only engine situation until mid-1963.

Stretching the basic Falcon structure, the Comet sedans were on a 114-inch wheelbase with an overall length of 194.9 inches. (Wagons shared the Falcon wheelbase.) Weights went up an average of 150 pounds.

Powering all this was the same 144-cube Falcon engine, which made for one of the slowest domestic cars on the market. *Motor Life* tested an automatic-equipped Comet four-door and found it took 27.5 seconds to reach 60 mph.

1961

Ford helped out the situation somewhat with the 1961 models by offering as an option a 170-ci version of the six. It boasted a 0.44-inch-longer stroke and a rating of 101 hp at 4400 rpm.

However the answer for future Falcon (and Comet) performance was not in power packs nor displacement increases for the six, but in V-8 power. Stroppe broke ground in 1960 when he installed a bored-out 312 Y-block V-8 in a Falcon four-door sedan. The 320-cube engine fit with a cut in the front shock towers, the driveline was converted to standard Ford, with a 1959 rear axle and three-speed Cruise-O-Matic.

The combination worked well and was given to William Clay Ford, who was then vice president in charge of styling. It went back to Dearborn and was used as his personal car.

1962

However, the use of a dated engine design (it had come out in the 1954 models) was not the best answer. It was not until Ford introduced its lightweight small-block in the 1962 model Fairlane and Mercury Meteor intermediates, that the right powerplant was available.

Of thin-wall cast-iron construction, the Fairlane V-8 had a meager 221 ci and rating of 145 hp at 4400 rpm. However, it didn't take the performance set long to realize that Ford finally had an answer to Chevy's famed small-block V-8. The Ford engine's dry weight was 470 pounds, compared to 356 for the 170 six. There was plenty of room to grow, and grow it did.

Smaller than the Y-block, the new small-block had little trouble easing between Falcon's front shock towers.

Meanwhile, the street Falcons had changed little. To counteract Chevrolet's highly successful Corvair Monza 900 coupe, a bucket-seat Falcon two-door came out in spring of 1961, dubbed the Futura. Comet got a similar addition, the S-22. However, the six-in-a-row situation under the hood had not changed.

While model year 1962 was rather dull for production Falcons (a formal-roofed Sports Futura was added mid-year, as was an English Ford four-speed manual transmission option), it was far from the case for the semi-official experimental versions of the compact. There were "secret" engineering V-8 Falcons running around Dearborn, but a couple of Holman & Moody Falcons made quite a splash with the public.

Holman & Moody was now a contractor for Ford, with the performance program back on track, and when the new V-8 came out, Holman & Moody wasted little time putting it to use. The V-8 was installed in a beefed-up Falcon two-door sedan, which was to be raced in the March 1962 Sebring twelve-hour sports car race.

Dubbed the Challenger I, it featured a bored-out version of the 221, at 243.968 cubic inches. This was just below the 244-cube (four-liter) limit for the prototype class. A four-barrel carb, special cam, 10.5:1 compression ratio, valve and ignition alterations further improved the output while a Borg-Warner T-10 four-speed manual transmission handled the horses and a modified big-Ford driveline got it to the ground. Ford front springs and a set of 1953 Ford leafs did the job in the back.

Painted red, white and blue (from the bottom up) it carried the number nine and was piloted by NASCAR ace Marvin Panch and sports car pilot Jocko Maggia-

Falcon's mid-year addition for 1962 came in the form of the Sports Futura two-door sedan. Features included a formal-type roof and new wheel covers. The first performance production Falcons were still a year away.

como. They went on to finish second in class (behind Jim Hall's Chaparral), but it was no great triumph as the team placed thirty-fourth out of thirty-five cars running at the finish, completing 107 laps to the winning Ferrari's 206. The Challenger wasn't all that slow, but it spent part of the race in the pits while the crew changed the heads.

Challenger I was followed by Challenger II, which was for the street. This one got the 260-cube V-8, which Ford brought out as a mid-year 1962 option for the Fairlane. It had a 0.3-inch larger bore than the 221.

By this time, the Carroll Shelby Cobra project was well underway. He put a 260 into an AC sports car, and his version was putting out that many (260!) horsepower, compared to the advertised 164 in the mild Fairlane 260.

Holman & Moody used the Cobra V-8 to power Challenger II, but the most noticeable feature was a three-inch-channel job resulting in a low-looking Falcon.

Road & Track tested Challenger II and found it a willing performer, doing 0-60 in 8.5 seconds, the quarter in sixteen seconds flat and with a top speed of 123 mph. Also noted were the number of rattles and imperfections in the handmade car.

A prototype fastback, Challenger III, was constructed by Holman & Moody, with the three-inch channel and a three-inch roof chop. There was talk of producing one hundred to qualify for production sports car events, but by then the Ford factory was well along with its own sporting Falcon V-8, the Sprint, which was due during the 1963 model run.

There were other small-V-8-powered Falcons around, including a Cobra-equipped Futura done by Andy Hotton of Dearborn Steel Tubing Company, who also was involved in special projects for Ford.

1963

However, with economy no longer the catchword in the compact class, and performance very much back in vogue for full-sized cars, Ford was almost forced to release the Falcon V-8 to the public.

The facelifted 1963 models at fall intro time were highlighted by Falcon's first convertible, a sporty-looking job, available in both the Futura and Sports Futura versions.

When the mid-year models came on stream, Falcon was in the performance compact business. Joining the bandwagon was a new body style, a two-door hardtop; a new engine, the 260; and a new performance option, the Sprint.

The new hardtop design was just what the Falcon needed to become a really sporty-looking package. Featuring swept-back rear pillars, at 53.2 inches high, it was 1.3 inches lower than the sedans. It was available in three versions, the Futura with a bench front seat, the Futura Sports Coupe with bucket-type front seats and the Sprint option, which featured front buckets and other items we'll look at shortly.

Also entering mid-year was the first production Falcon V-8, the 260, which came in mild form with a two-barrel carb, 8.7:1 compression ratio and a rating of 164 horses at 4400 rpm. In other models it was dubbed the Challenger 260 V-8, while in Sprints it was tabbed the Sprint 260 V-8.

Falcon's entry in the 1963 compact car performance race was the Sprint-optioned two-door hardtop. Standard came a 260 V-8, along with wire wheel covers, Sprint and V-8 emblems on the front fender, dash-mounted tachometer and heavy-duty chassis and suspension.

While it took three-and-a-fraction model years to get a V-8 into the Falcon, it was still the first of the Big Three low-priced compacts to get one. Chevrolet Corvair was forever to have its rear-mounted six; the more conventional front-engine Chevy II compact came as a 1962 model with four- and six-cylinder power and wouldn't get a V-8 until the 1964 models; Plymouth Valiant and Dodge Dart would only have six-cylinder power until mid way through the 1964 run; and even pioneer Rambler American would have to wait until mid-model-year 1966 for its bent-eight.

Studebaker Lark did have a V-8 from its 1959 model year intro on and the GM senior compacts—Pontiac Tempest, Oldsmobile F-85 and Buick Special/Skylark—had V-8's from the beginning of their introductions as 1961 models. The Challenger V-8 was optional in all Falcon passenger cars.

Four-speed manual Borg-Warner T-10 transmissions were a $188 option with the V-8. This compared with a $90.10 tag on the English Ford unit that was a six-cylinder option. The sixes had a three-speed box standard. While V-8's got an all-synchro three-speed without extra charge. Fordomatic was optional for both sixes and V-8's.

The addition of the V-8 brought larger 7.00x13 tires, while 6.50x13 was the previous largest size available. Suspensions were beefed-up, brakes were ten-inch drums all around, compared to nine before; and rear axle ratio was 3.25:1, compared to a six range from 3.1 to 4.0:1.

The star of the mid-year introductions was the Sprint option, available for the Sports Futura two-door hardtop and convertible. The Sprint 260 V-8 was standard. Though both Sprint and Challenger V-8's had identical specs, the Sprint version got less restrictive air cleaner and mufflers and came with chrome engine dress-up items, such as valve covers, air cleaner and oil filler cap, and its own decal on the covers.

Inside, front bucket-type seats, similar to the Futura Sports models, were standard along with a console between them, simulated wood-grain steering wheel and electric 6000-rpm tachometer mounted on top of the dash.

Falcon interiors for 1963 featured an elegant group of standard equipment and trim items, plus many popular options and accessories. Shown here is the standard electric tach mounted on instrument panel and simulated wood-grained sports-type steering wheel.

Falcon's first production convertible came in the 1963 lineup. The good-looking compact was offered from the start of the model year in Futura and Futura Sports versions, the latter with bucket seats. Later a Sprint option and V-8 engine were added.

Shadowing Falcon, Comet's first production convertible came at the start of the 1963 run. The S-22, shown here, was the sportiest with bucket-type seats, special trim and wheel covers. A hardtop joined the Comet line mid-year.

Sprint identification was on the instrument panel and at the leading edge of the front fenders with a chrome script and logo, the latter incorporating the V-8 symbol.

Exterior chrome was similar in design to that on the Futuras, with rocker panel trim added. Simulated wire wheel covers were standard and included simulated knock-off hubs.

Both Sprint hardtops and convertibles used the convertible version of the unitized chassis with torque boxes welded in for strength.

Despite all the neat stuff, the stock Sprint was no ball of fire. With a curb weight of 3,080 pounds for a convertible and 164 horses to pull it around, the Sprint had an 18.78-pound load for each horsepower. *Car and Driver* found a four-speed Sprint soft top could do 0-60 in only 12.1 seconds and cover the quarter in eighteen seconds, hitting the traps at 73 mph. While better than the 144- and 170-cube sixes, the stock Sprint was still well out of the major leagues in straight-line performance.

However, the hop-up potential of Ford's small-block V-8 was well-known both inside and outside of the company. And it would be modified Falcons that would establish a name in racing for the once plain-Jane compact.

Ford was into the performance thing as never before in the 1963 model year, and in order to introduce its 1963½ offerings to the press in a memorable way, Benson Ford took a jet full of key press people over to Monte Carlo for the finish of the famed Monte Carlo Rally in January.

Part of the plan was to enter a trio of Falcon Sprints in the rally, do well, of course, and use the whole thing to bolster the newly created Falcon image.

As it turned out, the Falcons got solidly beaten, but they did make enough headlines that most of the press was convinced that the elephant did jump the fence.

European rally competition is different from that in the U.S. and Canada. North American rallying is more a game of navigating skill, over there it's more an all-out race.

The trio of Sprint hardtops was prepared by Holman & Moody and shipped to Europe. The 260-horse Cobra 260-cid V-8 was used, as was a close-ratio four-speed gearbox with gears similar to those in the Corvette competition unit, a narrowed limited-slip Galaxie 4.51:1 rear axle, five-bolt fifteen-inch wheels, 11.5-inch Bendix/Dunlop front disc brakes and eleven-inch rear drums. Extra leafs were inserted in the rear springs.

Dunlop SP tires were to be used, but specially spiked units made by Fagersta Bruk of Sweden, added at the last minute, probably did more for the cars than any single component.

Three teams raced the cars. The combo that made the most news was that of ice-racer Bo Ljungfeldt and Gunnar Haggbom of Sweden. The other two were Anne Hall and Margaret McKenzie, and Peter Jopp and Trant Jarman.

The rally started from several places in Europe with the final segment covering common ground into Monte Carlo. The Ford team started from Monte Carlo, ran into bad weather and never was able to make up for it, as the number of penalty points assessed was too great.

However, there were six special stages, where the cars were timed in what amounted to be an all-out race. For the first time in the thirty-two-year history of the rally one car won all six; it was the Falcon of Ljungfeldt (who did all the driving).

Ford advertising and promotional people chose to play up that and the fact Ford did win its class for large-engine cars (rather than the point that Jopp and Jarman did the best) coming in thirty-fifth overall, eight places ahead of the charging Ljungfeldt. The women never did make the finish line. The rally was taken by Erik Carlsson in a two-stroke three-cylinder Saab 96, which Ford would rather forget.

At the start of the 1963 model year, the sportiest closed Falcon was the Futura Sports sedan, which was much like the 1962 model. Its place in the spotlight wouldn't last long, as the 1963½ hardtops would steal the show.

However the powerful V-8 Falcons did leave an impression with the fans and press. "The Falcons put on a fine display of power, sounding like fifty men tearing fifty phone books," said Henry Manney III in *Road & Track*.

The team was disbanded after Monte Carlo, but the cars were entered in events during the rest of the year, scoring well and even winning a few.

Production Falcons were also entered in rally events in North America, with a factory-backed effort taking the manufacturer's crown in the Shell 4000 Trans-Canada Rally, even though a Chevy II won.

The motoring press was able to drive the Monte Carlo Falcons and *Car and Driver* put one through the mill, alongside the aforementioned Sprint convertible. The magazine tester noted that the rally car could hit sixty in 7.5 seconds and tour the quarter in 16.8 seconds with a speed of 85 mph.

Mercury Comet also got many of the mid-1963 goodies that Falcon did. Its hardtop, which looked less well proportioned with its greater length, was dubbed the Sportster. It also got the 260 V-8, but had no special performance series like the Sprint. Promotion of the Comet as a performance compact wouldn't come until the 1964 model year.

1964

Speaking of the next model year, it was time for Falcon to get its first major face-lift. All sheetmetal from the beltline down was new, but not necessarily better-looking.

A spear started with pointed fender tips next to the headlights and gradually broadened as it ran to the back of the car, which featured the traditional bull's-eye taillights. Though the 1964 models looked bigger than the earlier ones, overall length was up only a half-inch and width increased by one inch. Weight went up about sixty pounds.

Wagons received an option of the 200-cubic-inch six, but overall mechanical changes were few. A few Falcons got out the door with the 289-cube version

One of the factory 1964 Falcon Sprints made its way through a narrow passage in the 1964 Monte Carlo Rally. The cars were specially constructed with fiberglass body panels and special high performance engines.

of the small-block under the hood, even though it was not listed as a regular production option.

Hardtop and convertible models were again available in Futura, Futura Sports and Sprint option configurations. Sprint equipment remained much the same as the previous year, but some vehicles with bench front seats instead of buckets were produced.

Monte Carlo was again the target for the Ford rally effort and a new batch of Sprints were prepared, with eight cars being entered.

The 1964 models were further from stock than the 1963's. Fiberglass body panels were used in place of steel to bring the weight down with hoods, trunk lids, front fenders and doors all being switched.

Power for the rally cars came from a 289 with a pair of two-barrel Carter carbs, a 10:1 compression ratio and a rating of 285 hp at 5500 rpm. Other chassis and driveline components were similar to the 1963's.

Car and Driver found the 1964 rally cars to be the fastest yet, with a six-second 0-60 trip and a 14.2-second run in the quarter.

The eight Falcon teams were among 299 starters in the event, which again originated in several cities. Four Falcons started from Paris and four from Oslo, Norway. Among them was Ljungfeldt, who again would be the star performer.

The publicity Falcon got in 1963 prompted Chrysler to enter its new Valiant V-8, in 1964, which competed in the Over 2500 cc GT class. That caused Ford to split its entries between the Over 2500 Touring and Over 2500 GT classes, so Valiant wouldn't be unchallenged.

New rules gave small-engine machines an advantage, but the lack of snow and the year's experience proved to help the Falcons considerably. All Falcons made it to Monte Carlo (though one crashed in the final leg there) and Ljungfeldt again was the fastest on all of the timed stages. However, he was tied on one by an Austin Mini-Cooper S driven by Paddy Hopkirk with Henry Liddon aboard. Using the small car factoring to his favor, Hopkirk was the overall winner.

Still sharing body and mechanical components with Falcon in 1964 was the Ranchero. For a while there was also a sedan delivery model available. Ranchero kept its Falcon ties through the 1966 model year, after which it was based on the intermediate Fairlane.

Sportiest of the 1964 Falcons was the Sprint-optioned Futura Sports convertible, shown here with the top down. Included in the package was the 260 V-8, wire wheel covers, tachometer on top of the dash and heavy-duty undersides. Falcon sprints were popular until the Mustang came along later in the model year.

Ljungfeldt however did much better than the year before and finished second overall and first in the Over 2500 Touring class. The Valiant threat never materialized and Anne Hall took the GT class win for Falcon.

The lightweight Falcons were later ruled illegal, but it didn't make much difference to Ford, for after Monte Carlo it would be concentrating on Mustang anyway.

The rally Falcons were later raced off and on with no major successes, but the fiberglass panels were utilized for another form of competition—drag racing. Ford had a batch of lightweight 427-powered Fairlanes constructed for super stock drag competition, known as Thunderbolts, and also arranged for some Falcons to be made for factory experimental class racing.

They were built by Dearborn Steel & Tubing and had 427 High Risers stuffed under their fiberglass hoods. They did well. One was driven by Phil Bonner to stock eliminator honors in the American Hot Rod Association World Championships in September of 1964 at the Green Valley Raceway in Ft. Worth, Texas. Bonner went through a field of the sixteen fastest stocks, blowing away several hot MoPars including Gene Snow's Plymouth Hemi in the final round. Bonner turned an 11.35/120.68 to take the honors.

Unlike the funny cars which were coming into use, the Falcons did not have greatly altered wheelbases. They did have modified fiberglass body panels, including the hood, to accommodate the high-rise intake manifold, fiberglass bumpers and air intakes similar to the Thunderbolts. Production was not high enough to qualify for production superstock classes, so they had to run in factory experimental competition.

With Mercury's intermediate Meteor gone, it concentrated on the Comet for the 1964 model year. It was restyled; though many elements were used, they seemed to blend into an attractive package. The popular Caliente hardtop is shown. It was joined mid-year by a Cyclone performance model.

Two of the factory-backed 427-powered 1964 Mercury Comets are shown in drag strip battle that year. Bill Shrewsberry (outside) goes against Don Nicholson. The success of the Comets prompted Mercury to shift its racing program from stock to drag competition.

Falcons were used in drag racing besides the A/FX examples. Skip Weld used a 1964 two-door sedan, a healthy 289 and non-stock hood scoop to do his thing on midwest quarter-miles in the late 1960's.

1965

Falcon lightweights were also built in 1965 trim, but this time Ford's single-overhead-cam 427 was used in place of the overhead-valve High Riser.

While the Falcon was just getting its performance game plan going, it was being undercut by Ford Division itself, as the new Mustang sporty compact, introduced as a 1965 model in April of 1964, would be the performance standard bearer for nearly all future compact car projects for several years.

Even though the Mustang was largely based on Falcon chassis and mechanical features, its unique styling and aggressive marketing made the Falcon look dated and boxy. Ford performance promotion and racing programs were quickly shifted to the Mustang.

Evidence of the Falcon's shift away from sports and performance orientation could be found when the 1965 models appeared. Gone were the Futura Sports models. The bucket-type front seats were now an option on the Futura hardtop and convertible, as was the Sprint.

However, the Sprint package contained only a few of the items that were included with previous Sprints. The 1965 option meant a V-8 engine, fender emblems, bucket seats and a console on the convertible. If you wanted to go real sporty, then you were advised to go Mustang.

On the outside, the 1965 Falcon was marked with a new grille and new side and rear trim, but the biggest changes were in the mechanical department.

Dropped were the 144 six and 260 V-8. A 105-hp version of the 170-cid six was standard, with the Fairlane 200-cid six (at 120 hp) and 289-cid Challenger V-8 as options. The latter was available in mild form, with a two-barrel carb and rating of 200 hp.

Transmission selection was also changed. Three-speed manual units were standard across the board, with the V-8's getting all gears synchronized. Three-

In order to take on the MoPars, a few A/FX Falcons were built for drag racing. They were very fast, as the 1964 models had a 427 High Riser for power and the 1965's sohc 427's. Lightweight body panels, altered wheelbase (two inches) and modified suspension were some of the features. One of the most successful drivers was Phil Bonner, shown here in the Holman & Moody 1965 Falcon running against Dave Strickler's A/FX Dodge Coronet at the Lions Drag Strip.

Back for its final year on its original chassis was the 1965 Falcon, shown in Futura hardtop form. Among the changes that year were a new grille, side trim, dashboard and rear trim. There also were changes in the powertrain with the 289 being a regular production option for the first time. The car pictured has the optional wire wheel covers that used to be part of the Sprint package.

speed Cruise-O-Matic replaced the two-speed Fordomatic as an option for all engines. The four-speed manual box was only optional with the V-8. The British four-speed unit was dropped.

Also new was the use of fourteen-inch wheels and tires. All V-8 equipped models came with them, as did six-cylinder wagons. They were optional for other sixes. The 6.45x14 four-ply rated tires were standard on V-8 sedans, convertibles and hardtops, while 6.95x14's were an option. Five-bolt wheels were standard on V-8 cars, while sixes got four-bolt wheels.

Falcon Sprint popularity waxed and waned in inverse relation to the Mustang. Only available for part of the 1963 model year, 15,081 Sprints were produced. During the 1964 model run, 18,108 Sprints were called for. In 1965, the number of Sprint-optioned Falcons dropped to 3,106.

1966-1970

After 1965, Falcon was continued as a shortened version of the Fairlane. Two- and four-door sedans, plus wagons were offered. The Ranchero pickup was Falcon-based for 1966, but from 1967 on used Fairlane trim.

Sedan wheelbase went up to 111 inches from 109.5 while wagons and Rancheros went to 113, a dimension that was shared with Fairlane. Sedan overall length grew to 184.3 inches from 181.6 and width was now 73.2 inches, 1.6 more than before. Weight grew 150 pounds.

Gone were the hardtops, convertibles and Sprint options. The sporty model was the Futura Sports Coupe, which came with bucket-type seats standard.

Billed as the "World's Durability Champion," Comet was promoted through endurance runs, rallies and drag racing. A 1965 Caliente hardtop is shown, complete with optional wire wheel covers. It was the last year for the Comet in the compact car field for a while. Starting in 1966 the name was put on intermediate-sized cars.

Comet had no Mustang sporty compact to sell for 1965, so it made do with the Cyclone hardtop, which came with a 289, special grille, trim and wheel covers that looked like chromed wheels.

Engine lineup remained the same as the previous year, as did the choice of transmissions.

Falcon had now fully reverted back to the way it was. Performance promotion and racing activities centered around other models, while the Falcon provided rather basic transportation.

The 1966-style models remained relatively unchanged during their production run, which lasted to the end of calendar year 1969.

Smaller Maverick sedans came out in spring of 1969 as 1970 models and spelled doom for the larger Falcon. Ford announced 1970 Falcons in the fall of 1969, but only built them until the end of calendar 1969.

There was a 1970½ version, with Falcon nameplates placed on low-priced models of the intermediate Fairlane. Ironically, these were the highest-performance Falcons of all, for they could be optioned with Ford's 429-cube V-8 with a 370-hp advertised rating. However, with its economy theme, few were likely so equipped.

The Falcon models were absorbed in the Torino series before the 1970 model year was over, ending an eleven-year life span for the nameplate.

Falcon's counterpart in the Mercury lineup, Comet, fared quite differently. Part of the reason for the performance emphasis that would be placed on Comet was its surroundings.

Contrasting the old and new Falcon bodies was this 1966 A/FX match at the Amarillo (Texas) Dragway. Tom McNeely (inside) wheels his 1965 Falcon against the McLellan Brothers 1966 model. A/FX cars were approaching funny cars with altered wheelbases and set-back bodies.

Even Rancheros got into the drag racing act, as their mechanical components were the same as Falcons through 1966. This 1965 Ranchero was driven by Dave Pinta at the Great Lakes Dragaway in Union Grove, Wisconsin.

Starting with the 1966 models, Falcon used a shortened Fairlane chassis for its sedans. Wagons used the same chassis. Sportiest version was the Futura Sports Coupe, shown here, with bucket seats and special trim. Engine selection was mild with the performance tricks being done in the Mustang line.

Mercury brought out its intermediate Meteor as a 1962 model, which shared much with the Ford Fairlane, including basic body, chassis and engines. Since the Comet was only 2.5 inches shorter in wheelbase and sold for a lower price, there was some confusion as to what the Meteor was, as intermediate-sized cars were rare in 1962.

There was no such confusion in the Ford stable however, as Falcon was a smaller car than the Comet and the Fairlane fit nicely between the Falcon and full-sized Ford.

Meteor was dropped after the 1963 model run, leaving a restyled Comet to carry on. Also, there would be no Mustang to dilute the Comet during the 1964 model run, so Comet would play many roles for Mercury; compact, sporty compact and intermediate.

Just like Ford, Mercury got the go-ahead to get into racing and performance promotion, and the first venture was to enter the full-sized Mercury in late model stock car racing.

Though the parts were right, which included the big 427 V-8, the package was a bit too large for the highly competitive NASCAR scene in 1963. Smaller, lighter Fords, Plymouths, Dodges and Chevrolets all combined to make the season a dismal one for Mercury.

Emphasis shifted to the 1964 Comet. Like Falcon, Comet got new sheetmetal from the beltline down for 1964, and while being of rather complicated persuasion, the combination of chrome, sculpturing and conflicting angles seemed to give the car more character than the rather nondescript earlier models.

A new Caliente series at the top of the line contained hardtop and convertible models. Bucket seats were optional. Also optional for all 1964 models was the 289 V-8. It came with a four-barrel carb, but its 9:1 compression ratio meant a taste for regular gasoline. Transmission availability was similar to that for Falcon V-8's, with three-speed all-synchro standard and four-speed manual optional. Also optional was Multi-Drive Merc-O-Matic, which was like the Ford three-speed Cruise-O-Matic.

Announced in January of 1964 was the Comet Cyclone, Mercury's answer to the Sprint. The 289 was standard and came with a chrome dress-up package. Bucket seats and a console were also standard, as was a dash-mounted tachometer, simulated wood steering wheel and Cyclone fender emblems. Showing it could simulate with the best of them, the Cyclone got wheelcovers which looked like chrome wheels. Chrome lug nuts came through holes in the covers and the whole setup did look fairly realistic.

Later in the model year, some Comets were made with the 271-hp high-performance version of the 289.

To convince the motoring public that the Comet was a performer, a group of 1964 Comets was sent to Daytona International Speedway before new model

Falcons changed little during the 1966 through early 1970 model years. A 1969 Futura Sports Coupe is shown. Production of 1970 models ceased at the end of 1969 and the Falcon name appeared briefly on an intermediate car later in the model year.

introduction and run for 100,000 miles. They averaged 105 mph for the jaunt and set many records. The feat let Mercury advertise the Comet as the "World's 100,000-Mile Durability Champion."

Comets were also entered in the East African Safari Rally and out of ten cars on hand, two finished. But then, only twenty-one of the ninty-four starters were able to complete the rally, so despite the fact the Comets finished eighteenth and twenty-first, more publicity was gained for Comet.

Drag racing fans came to know the Comets quite well, as a number of light-weight, 427-powered Comets were constructed for the 1964 season. With factory backing they scored many major wins in factory experimental competition.

Comets got a slight facelift for the 1965 run, with vertical dual headlights, instead of horizontal, being the most visible change. Two- and four-barrel versions of the 289 topped the list, respectively listed at 200 and 225 hp.

Trying to top its 100,000 mile 1964 promo, Mercury sent a trio of Comets on a 16,247 mile excursion from the tip of Cape Horn in Ushuaia, Argentina, to Fairbanks, Alaska. They made it in forty days without a failure of any kind and the journey was hyped for the duration of the model year.

The A/FX drag Comets were back for 1965, this time with overhead-cam 427's, and again did well, but got heavy competition from Mustangs with the same power-plant.

Since Comets had more in common with the intermediate-sized cars than the compacts, the 1966 redesign saw the Comet become an intermediate, sharing size and basics with the Fairlane. The Comet name faded during the 1960's, being replaced by the Montego tag. However, it was dusted off for the 1971 model year and put on Mercury's version of the Ford Maverick, where it remained through the 1977 model year.

Ford's SOHC "Cammer"

Early in 1964, Chrysler set the performance world on its ear with the Hemi. The powerful 426-cube engine produced more power at the time than any other production domestic passenger car engine. The Hemi premiered at Daytona in February of 1964 and took the first three places in the famed Daytona 500. Everyone noticed.

Chrysler didn't have to design an all-new engine. It just took its big-block and adapted hemispherical heads to it, similar to those it used on its FirePower engine in the 1950's.

Ford's answer to the Hemi, the single-overhead-cam (sohc) 427 was done in much the same fashion. It adapted its hefty 427 block and designed new heads with a chain-driven cam on top of each. While the chain drive and gearing for it was somewhat complicated, the result was a production-based engine that put out more power than the Hemi.

The only problem was that Chrysler got away with putting the Hemi in its NASCAR stock cars and Ford didn't. Eventually a street Hemi came out and it became a regular production option, but that took a year and a half after the engine was legal for racing.

Ford threatened a production street sohc and played with the idea, but it never came to pass. The initial target for the sohc was NASCAR stock car racing, but as we see in this chapter, that never came to pass either.

As it turned out, the playground for the sohc would be drag racing. The Hemi was the king of AA fuel drag racing in the 1960's, but it wasn't the new one being used in stock competition. Rather it was the old FirePower, which was last installed in 1958 Chryslers.

In the 1965 season, Connie Kalitta, known as the "Bounty Hunter," was given an sohc for his rail to see what he could do. At first it wasn't much; but gradually Kalitta and later Pete Robinson got the bugs out of the sohc. They started winning and Kalitta spent a couple of seasons as one of the hottest rail pilots in the country. Things jelled in late 1966.

Also, the sohc went into Ford's 1965 factory experimental Mustangs, Falcons, Galaxies and an occasional Fairlane. Mercury Comets also got them. That continued for 1966. They did quite well.

While the sohc could have been as long-lived as the Hemi, the basic design which still dominates AA fuel dragster competition today, Ford lost interest around 1967 or 1968, as it was developing a new engine based on the new 429 block. The Boss 429 evolved, put out lots of power and was cheaper to produce. It was legalized for NASCAR and found some acceptance in drag racing.

The Boss may have been the engine of the future, but for awesome power and even looks, it never quite equaled the sohc, known today as the "Cammer."

It took awhile, but Connie Kalitta made the Ford sohc 427 a winner in AA fuel drag racing, beating the Chrysler Hemis.

Among the converts from Hemi to sohc Ford power was Don Prudhomme, shown in a 1967 photo. At 25 years old, he was already considered one of the top rail pilots at the time.

CHAPTER FOUR

Totally Performing Big Fords And Mercurys

In the first few years after World War II, there more or less was a direct relationship between performance, size and price tag among the new domestic passenger cars. Cars like Cadillac, Lincoln and Packard had the highest horsepower ratings, top speed and price tags.

This order was broken in 1949 when a medium-priced car, the Oldsmobile 88, got a modern overhead-valve V-8 and gained a reputation as a hot performer. However, up through 1954, medium- and high-priced big cars dominated the performance statistics and racing events, like Hudson Hornet, Oldsmobile 88, Chrysler New Yorker, Lincoln and Cadillac.

Another drop in size and price for performance cars came about during the 1955 model run when Chevrolet and, later, Ford built, sold and raced performance cars that could beat most of the big ones.

The reign of the low-priced, full-sized car in the rough, tough performance world didn't quite last a decade, as they became victims of the formula by which they came to power. Lower-priced and smaller intermediate and compact cars dethroned the big 'uns in the wild mid-1960's.

The performance role of the big Fords was interrupted by the June 1957 Automobile Manufacturers Association ban on factory involvement in performance promotion and racing. At the time the ax fell, you could get a 312-ci V-8 with a McCulloch supercharger aboard and fairly realistic rating of 300 hp.

For 1958, Ford announced a new big-block, the FE engine series, in two versions, 332 and 352 ci. The latter also came with a 300-horse rating, but didn't match the hot 312 in actual output. Ford would carry the 300 rating for many years, despite a variety of engine sizes and tune.

Originally, the 352/300 came with solid lifters and did have some performance potential, but after adjusting to lead time and the AMA ban, they were replaced with hydraulic lifters and the compression ratio was cut from 10.2 to 9.6 early in the 1958 model run.

For 1959, Fords got all-new sheetmetal, which probably wouldn't have been all that noticed, except that Ford's main competitor, Chevrolet, went off the deep end in the style department with radical fins, taillights and front-end lines.

Ford outsold Chevrolet for most of the model year, which surprised Ford as much as anyone, for it had its own radically styled car ready to go for the next model year.

Ford power for 1959 was still comprised of the 300-horse 352 on top, while the competition, especially Chevrolet and Pontiac, was busy adding new engine options at the high-performance end of the list. Ford's only ray of hope was the 430-ci Lincoln V-8 that was optional for the Thunderbird that year.

Ford had followed the AMA edict to the letter and was the only one among the Big Three to do so. When it became apparent that Ford was the only one on board the ship, a letter was sent to General Motors on April 27, 1959, stating that Ford was thinking of offering some performance options of its own and suggesting that the AMA agreement be reworked.

There was no answer.

With the country recovering from its 1958 recession and the sport of auto racing growing, with several superspeedways for stock car racing opening in the

Southeast and the sport of stock car drag racing getting organized and gaining in popularity, the performance market was just too big to ignore, especially if Ford was the only one doing the ignoring.

With non-factory-supported Thunderbirds holding their own in NASCAR Grand National and convertible races, Ford went to work to put its performance parts game back on track after two years of inactivity.

There was no immediate outbreak of performance cars or parts. A three-man group, headed by engineer Dave Evans, was formed to develop a performance program. Also on board were pioneer engine man Don Sullivan, whose work dated back to the early V-8's, and young chassis engineer John Cowley. Their guise was "law-enforcement parts development," something that had been used at Chevrolet, Pontiac and Dodge after the AMA crackdown.

1960

There would be no new engines or other exotic equipment like superchargers for the performance program at this stage of the operation and indeed Ford would be lucky to even get an engine setup in production, considering the short lead time before the 1960 models hit the market.

The only logical choice for a powerplant was the 352. It was close in size to Chevrolet's 348-cid "W" engine and Plymouth's top engine for 1959, the 361-cid version of the "B" block.

Some of the good stuff was right on the shelf for Ford. The mechanical lifters from the early 1958 models were used, as were pushrods from the soon to be introduced Falcon six. The heads were redesigned so the combustion chambers got a 10.6:1 squeeze. A hot cam helped the cause.

New was an aluminum intake manifold that saved forty pounds over the cast-iron unit. On top of it went a 540-cubic-foot-per-minute Holley four-barrel carburetor. A new set of cast-iron exhaust headers did the extraction job. A cast, nodular iron crankshaft kept all in place underneath and a dual-point ignition supplied enough spark.

Tests showed the Cruise-O-Matic was not up to the task of transmitting the power from the new high-performance 352, so the Borg-Warner T-85 three-speed manual or overdrive would be the only choices.

After dynamometer testing, the new Interceptor engine would be rated at 360 hp at a healthy 6000 rpm. This compared to the mild 4600 at which the alleged 300-hp was developed. Torque worked out to 380 pounds-feet at 3400.

One of the highlights for Ford in the late 1950's was short-track driver Fred Lorenzen's (#28) winning both the 1958 and 1959 USAC stock car championships in a 1958 Ford. He is shown here in a 1958 event leading 1957 Chevy driver Tom Pistone. Lorenzen went on to become a NASCAR superstar for Ford in the 1960's.

Ford styling for 1959 was a hit with the new car buyers. It was helped by a mid-year addition, the Galaxie series, shown in four-door hardtop form. In the background is the car that inspired the Galaxie roof, the 1959 Thunderbird.

With local testing done, it was time for bigger things. In August of 1959, the prototype car was taken to Daytona International Speedway for some laps on the big 2.5-mile oval.

It was natural to think Ford's former factory reps, Holman and Moody, would be hired, but that would be too obvious, so the lesser known Wood Brothers of Stuart, Virginia, were called. Hired to drive was Cotton Owens, who made a name for himself racing Pontiacs in NASCAR.

The car worked right out of the box and Owens turned five laps at 145.5 mph and forty at 142, which was competitive with the 1959 machinery. A prototype was then run at Ford's test track in Romeo, Michigan, and hit 152.6 mph.

Ford's performance work wasn't exactly done in a shroud of secrecy, once Evans felt he had a winner. The motoring press was let in on the project.

Pioneer postwar auto writer Tom McCahill of *Mechanix Illustrated* got to drive one of the preproduction 360-hp Fords and in his usual style, described the difference between it and the regular 1960 Ford he was testing. "It's like saying that Eddie Arcaro and Wilt The Stilt are identical physical specimens since they both have ears," he noted.

Motor Life tested an early version and found it capable of a 0-60 sprint in 7.1 seconds. The 300-horse jobs were averaging ten seconds and up for the same task.

"It took several years, but we think Ford has the right answer for 1960," said Ray Brock in the respected *Hot Rod* magazine.

Conservative styling may have been the hot setup for Ford in the 1959 model year, but it was to be a one-year ordeal, as an all-new car was ready for the 1960 model chase. While not quite as radical as the 1959 Chevrolet, it was the biggest change for the standard Ford since the 1949 models.

While the outside body may have looked stock in the early 1960's, in order to be successful in circle-track stock car racing, the frame had to be beefed-up considerably. Shown in the Zecol-Lubaid shop is a Ford frame ready for stock car competition. Note the stronger front suspension with heavy control arms, air bags and towers made for the twin shocks. The rear also is rearranged with four shocks and bracing. Heavy brakes, completely rewelded frame rails and other bracing are all part of the package.

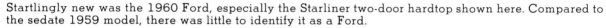

Startlingly new was the 1960 Ford, especially the Starliner two-door hardtop shown here. Compared to the sedate 1959 model, there was little to identify it as a Ford.

If the "longer, lower, wider" theme ever described a car, it was the 1960 Ford. Wheelbase went up an inch from 118 to 119 inches, but the overall length grew 5.7 inches to 213.7. Width ballooned from 76.6 inches to 81.5. The latter figure was illegal for highway operation by a passenger car in most states, without proper marker lights. The maximum allowable width was eighty inches. Ford wasn't the only one over the limit, Edsel (which shared the same body), Mercury, Chevrolet and Oldsmobile all overstepped their bounds with their 1960 models.

The manufacturers agreed to make their next models narrower and the law agreed to look the other way. Thus, the 1960 Fords were the widest passenger cars ever made by the division.

Height varied per model, but two-door hardtops for example dropped from fifty-six inches loaded, to 54.5.

While Pontiac was making all kinds of noise about its Wide-Track suspension, Ford did a bit of expanding on its own. Front tread grew from fifty-nine to sixty-one inches and rear from 56.4 to sixty. The numbers were still short of Pontiac's sixty-four however.

Though the basic theme was the same, styling for the 1960 Fords was carried off with varying degrees of success. At the top of the list was a sleek two-door hardtop, the Starliner.

Starting with the nonwraparound windshield (Ford's first since its 1954 models), the roofline sleekly flowed back to a contoured rear deck. The rear pillar was tastefully decorated with a trio of chrome doo-dads and on two-tone cars, the color line ran along the pillar to the windshield.

Good looks weren't the only result of Starliner styling, it would also help the aerodynamic cause at places like Daytona.

Sharing the spotlight with the Starliner was the Sunliner convertible. Top down, it showed off the fairly clean lines of the 1960 models to the fullest. Not coming off bad were the Galaxie sedans and four-door hardtop, which had formal rear roof pillars. Somewhat less coordinated were the lower-priced Fairlane and Fairlane 500 sedans which had thinner rear pillars and large wraparound rear windows.

From the beltline down, all cars shared the same styling. Headlights were in the full-width grille, which was flanked by ridges that sprouted from the bumper, went over the top of the fender and continued straight to the back, where they curved into the rear deck. Ford's familiar bull's-eye taillights took the year off, replaced by single semicircular jobs that were at either end of an indented full-width panel.

The whole package looked like no Ford before it—and at the time, if you would have removed the nameplates, few would have known what make it was.

Galaxie 500, Sunliner, Starliner and Country Sedan station wagon side trim was a single spear of chrome, extending from the rear bumper about two thirds of the length of the car. Behind the rear wheel, textured anodized-aluminum filled-in the open space.

Coming off a year with the sales lead, having a sleek all-new car to sell and new vigor in the performance field, Ford seemingly had it made for 1960. However, things just didn't work out that way.

Also introduced for 1960 was the compact Ford Falcon. Its conventional styling and engineering caught on with the buyers, who put it in first place in the

The 1960 Fords were the widest ever at 81.5 inches, and looked it in this view of the Sunliner convertible. Gone for a year were the bull's-eye taillights. Despite their good looks, 1960 Fords didn't sell as well as the 1959's.

Pit stops seemed somewhat more leisurely than the rapid ones today. Curtis Turner gets his 1960 Ford Starliner serviced at the 1960 Daytona 500. The sleek roof lines helped the Fords on the superspeedways that year.

compact car sales race. Falcon sales cut into those of the lower-priced, big Ford line. And to make things worse, there was resistance to the new styling, similar to what Chevy experienced with the 1959 models.

When the production lines stopped, 911,034 1960-model big Fords had rolled off the line from U.S. plants, down a whopping 483,647 from the previous model year. Even with Falcon added, the 1,346,710 total fell short of the regular 1959 Ford run of 1,394,681.

Fortunately, the situation on the racetracks was somewhat more encouraging. When the word of Ford's tests leaked out, interest was high among NASCAR independent competitors. Holman & Moody helped things along by offering a race-ready Starliner for $4,995, which was cheaper than its 1959 Thunderbird racer.

Only one 1959 regular Ford was in the Daytona 500 field, but come February 1960 it was a different story. Ford was the most popular car in the field with thirty-two entered and twenty-four in the show.

However, going fast on a few test laps and doing it for five hundred miles are worlds apart—so the Ford fellows found out.

In the two hundred-mile qualifying races, where all-out speed is required, the Pontiacs and Chevrolets ruled the roost. They were better prepared and were a lot further from stock than the Fords. Fireball Roberts, driving for Smokey Yunick, won the first in a 1960 Pontiac at a record 151.556 mph. The best performance for a Ford was turned in by two-time United States Auto Club champion Fred Lorenzen of Elmhurst, Illinois, who finished third, behind Cotton Owens's Pontiac.

Jack Smith claimed the second century grind, making a sweep for Pontiac. The first 1960 Ford didn't cross the finish line until thirteenth, wheeled by Johnny Sudderth.

When the 500 started, so did the Fords's problems. Engines expired in quick fashion with four of the first five cars out of the race with blown engines.

No Ford led the race and when it was over Junior Johnson, who said he hadn't been ready in time for last year's race, was the winner in a 1959 Chevrolet. Bobby Johns was second in a year-old Pontiac, followed by the 1960 Plymouths of Richard Petty and his father Lee, the 1959 winner. Johnny Allen took fifth in a 1960 Chevy.

The first Fords across the finish line were sixth-place Ned Jarrett followed by the legendary Curtis Turner and struggling newcomer Lorenzen.

It was apparent more work would be needed to make the Fords competitive.

Another popular part of the February activities in Daytona was the Speed Trials on the beach. Even though the big races were now held at the impressive Daytona track, the beach would be used through the 1961 Speed Week activities.

Fords were plentiful in the Flying Mile runs for class five (305 to 361 ci), but were no match for the Chevy 348's. The fastest run was by Chevy driver Harry Perry with an average speed of 136.208 mph for his run in each direction, but his car tossed a fan belt, which meant disqualification. Fellow Chevy pilot Jim Rahm got the win with a 135.287 average. The Fords? The best for a Starliner was in the 132 range.

Actually, the 1960 season didn't turn out all that badly for Ford drivers in NASCAR Grand National racing as they won fifteen races, more than any other make. Chevrolet pilots were next with thirteen, followed by Plymouth with eight, Pontiac at seven and Dodge with one.

Fairlane sedans were popular stock cars in forms of racing other than NASCAR Grand National, where aerodynamics weren't all that important. Shown is a 1960 example driven that year in USAC by Norm Nelson who took it to the championship.

Stock car, drag and street racers all went for Ford's 1960 Interceptor 352 V-8, which was advertised at 360 hp. It is shown aboard a stock car from that year. Note the nonstock suspension modifications.

Ten of Ford's wins were scored by 1960 models, with two coming on superspeedways, an increasingly important statistic to the manufacturers, due to the news value and drawing power at the big tracks. In addition to Darlington (which opened in 1950) and Daytona, new 1.5-mile tracks opened near Atlanta, Georgia, and Charlotte, North Carolina, in 1960.

Weatherly won the Rebel 300 convertible race at Darlington on May 14 and Speedy Thompson was just that in the October 16 National 400 at Charlotte.

Despite producing high-performance engines and an increasing number of heavy-duty parts, Ford still was not in the business of backing racing teams in NASCAR during the 1960 season, but that would change.

When all the points were counted, Chevy driver Rex White won the NASCAR Grand National championship. Fortunately, Ford wasn't shut out in the driving title department in other areas of the country. In the USAC stock car circuit, which ran in the Midwest, Norm Nelson of Racine, Wisconsin, won three of the nine events on the schedule in a 1960 Ford Fairlane sedan entered by Zecol-Lubaid chemicals of Milwaukee, and took the series championship, his first. Fords accounted for eight of the nine wins, with Paul Goldsmith driving a Ray Nichels Pontiac to a single flag.

The efforts of the Zecol operation, headed by Mrs. Vanda Hurst and Bill Trainor, would not go unnoticed in future Ford racing plans. Neither would the performance of Nelson Stacy of Cincinnati, Ohio, who won the New Car driving championship of the Toledo, Ohio-based Midwest Association for Race Cars (MARC), also aboard a 1960 Ford Fairlane sedan.

In drag racing, Ford didn't win big, but Interceptor-powered cars did set some strip records, battling the tough Pontiacs and Chevrolets. The 360-hp engine was thrown into the super stock class due to its high-horsepower rating. That year Pontiac, with its 389-cid engine, had a top rating of 348 hp and the 348-cid Chevrolets topped out at 335.

Also, since production automatic Interceptors were not readily available, Ford could only compete in the classes for stick-shift cars.

One thing that hurt Ford performance efforts besides lack of experience was lack of cubic inches. In 1960 Pontiac had 389, Dodge and Plymouth 383 and only Chevy had fewer at 348. Rumors of larger engines loomed for 1961 and Ford thought it was ready for the battle.

1961

The FE engine went through three model years without a displacement increase. Considering the cars it was powering had gained a couple hundred pounds and the growing demand for performance, enlargement was in order.

New at the top of the Ford engine line for 1961 was the 390, arrived at by increasing the 352 bore 0.05 inch to 4.05 and the stroke 0.28 to 3.78.

Dubbed the Thunderbird 390 Special V-8, it was said to produce—you guessed it—300 hp, and at the same 4600 rpm as the 352. Torque did go up from 381 to 427 pounds-feet, still at 2800 rpm.

The 352 was still around, but only in two-barrel carb, regular-gas-burning form. Called the Thunderbird 352 Special V-8, it was only good for 220 hp. (In case you are wondering about the use of the Thunderbird name, Ford was very liberal with it when it came to V-8 engines and that included the lowly 292, which hadn't been in a Thunderbird since 1957 and was called the Thunderbird 292 V-8.)

Each year one race on the NASCAR Grand National circuit was run with open-top cars instead of hardtops, the Rebel 300 at Darlington International Raceway. Fred Lorenzen is shown starting on the pole in the May 6, 1961, event, driving a Holman & Moody 1961 Ford (#28). He is side-by-side with 1961 Pontiac driver Fireball Roberts. Also in formation are Joe Weatherly (#8-1961 Pontiac), Banjo Matthews (#94-1961 Ford) and Cotton Owens (#6-1961 Pontiac). Lorenzen won the race.

A more potent version of the 390 was available supposedly only to law enforcement agencies, the Interceptor V-8, rated at 330 hp at 5000 rpm. It had solid lifters, high-lift cam, header exhaust manifolds and other good things.

However, for the general public, the storm trooper was the Thunderbird 390 Super V-8. Suspicion that Ford was backing off a bit on advertising horsepower was aroused when a rating of 375 at 6000 was announced. Displacement went up thirty-eight cubes, yet only fifteen horses were gained.

In drag racing, class was determined by weight and advertised horsepower. The lower the advertised horses, the better chance you had, something Pontiac had learned many seasons ago.

Blocks used in the high-performance 1960 engines were checked for quality, as were the other components, but were basically the same as the regular 300-horse units.

For 1961 this situation changed with the high-performance 390 block being different from the standard edition. The precise quality control was still there, but the casting was different. The main bearing webs were cast thicker with reinforcement ribs, making the bottom end of the already strong Y-block FE engine even tougher. An extra oil-pressure relief valve was mounted in a boss, cast between the fourth and fifth web. Enlarged oil passages were also included in the casting, and blocks were dye-checked for minute cracks.

Other minor engine changes were made, but the heads and cam were retained from the previous model. To feed the new cubes, a new larger Holley carb and intake manifold with ten-percent-larger passages were added. The cast-iron exhaust headers from the 360 were retained and again exited into larger-than-stock dual exhausts.

Ordering a 375-hp engine in a new Ford in 1961 meant there were some things that came with it automatically, and some things that didn't come with it, no matter how much you pleaded. This policy would be continued on full-sized Fords for several years.

First, the 375 could be ordered in all models except station wagons. Second, if you liked automatic transmission, power steering or power brakes, you had to choose another powerplant, because the options were not available with the 375.

What you did get were heavy-duty springs and shocks, all around, three-inch-wide front brake drums instead of 2.5 on the standard models, fade-resistant linings, fifteen-inch wheels instead of fourteen, nylon cord tires with 6.70x15 standard (7.10x15 optional), a three-inch-diameter driveshaft instead of the 2.75-inch unit and a four-pinion differential in place of the two-pinion regular unit. Heavier U-joints were also included.

Once again the Borg-Warner T-85 three-speed transmission was standard with overdrive optional. Ratios for the three-speed unit were 2.37 in first and 1.51 in second. The axle ratio for that setup was 3.89:1. Overdrive gears were 2.49 in first, 1.59 in second and 0.72 in overdrive, all working through a standard 4.11:1 differential.

While production line choices were limited, Ford dealers carried gear sets in their parts departments from 3:1 to 5.83:1, which could take care of any situation from highway cruising to quarter-mile runs against the clock.

Awaiting shipment from Ford's Dallas, Texas, assembly plant was a 1961 Ford Galaxie 500 Club Victoria hardtop, which was new to the line that model year. Other 1961 Fords were in the background.

Both Starliners and Fairlane sedans were again used on the USAC stock car circuit in 1961. Jim Rathmann (#43) raced a 1961 Starliner for Zecol-Lubaid and Norm Nelson (#1) is in a 1961 Fairlane sedan. Both cars had factory support.

Technology was advancing rapidly in the early 1960's and the hot setup at the start of the season was usually just the first stage of the program. Hotter options were saved for later in the model year, for a couple of reasons.

First, racing sanctioning bodies usually had a time limit for announcing models, engines or options before their events. At the time it was forty-five days. So if you wanted something new to be legal for the Daytona 500 or NHRA Winternationals at Pomona, California, in February, you waited until December to put your cards on the table. This way you could use the latest developments and your competitors would not have time to react and be legal for the event.

While we're into racing sanctioning bodies, it should be noted that in the early 1960's, there were different engine requirements for circle track stock car racing and drag racing.

NASCAR, along with USAC and other stock car groups, required a single four-barrel carburetor on its engines, a situation that lasted until mid-way through the 1966 season.

Drag racing groups, notably the NHRA, American Hot Rod Association (AHRA) and NASCAR (which also sanctioned drags for awhile) let multiple carburetion be used, meaning three two-barrels, or two four-barrels.

As a result, factories usually announced their hottest engines in two forms, so as to be legal for both types of competition. Sanctioning bodies insisted that the setups used in stock competition had to be available to the public, although there were many violations of the rule over the years.

Now we know what's under the 1961 Fords, let's take a look at what's bolted to the chassis.

The same basic body shell was used as in the 1960 line, but all sheetmetal below the beltline was changed. Since the 1960 models, which had little identification from previous models, bombed with the public, the theme was back to some familiar features.

"Beautifully proportioned to the classic Ford look," was the advertising theme. Ford arranged to be honored for its efforts by Centro per L'Alta Moda Italiana, which was billed as an international fashion authority. The award was given for "functional expression of classic beauty."

The most noticeable feature of the 1961 Fords was a return to the traditional round taillights, accented by a small fin that extended forward to form the front door handle. The grille was rather nondescript, being full width with a center bar. The grille pattern was repeated in a panel between the taillights on top-line Galaxie models.

Galaxies had the chrome spear of 1960 extended to the front wheel well and a modified aluminum trim panel behind the rear wheels.

Once again the base Fairlane with less of everything, started things off, though it was minus the business coupe it had in 1960. Fairlane 500 was the next step up which, like the Fairlane, had a wraparound rear window.

At the top of the line was the Galaxie, consisting of two- and four-door sedans, a new formal-roofed two-door hardtop and a four-door hardtop.

In 1960, Ford was very vague about the Starliner and Sunliner's position in the Galaxie series. They did not carry Galaxie nameplates on the fenders, trunk or dash, but rather their respective Starliner and Sunliner logos. For 1961, at least the trunks got Galaxie plates.

Getting the bulk of the promotion for 1961 was the Galaxie Club Victoria, the formal-roofed hardtop. While the Starliner may have been just what the racer and performance buyer wanted, the more conservative lines of the Victoria were felt to be more in line with the tastes of the traditional Ford buyer.

Speed Trials were held on the sands of Daytona Beach, Florida, for the final time in 1961. Don White took a brand new 1961 Ford Starliner there, complete with 375-hp 390 and established a flying-mile class record there of 159.320 mph. The car was not a race car despite the decals of his USAC sponsor, but rather a stock passenger car with a roll bar welded in.

Indeed, when the final tally was in, 75,437 Victorias were called for from the factory, compared to 29,669 Starliners. The result no doubt played a role in Ford's decision to ax the Starliner from the 1962 model selection.

Even though the same basic body shell was used, a couple of key dimensions shrank on the 1961 Fords. Width went down to a legal 79.9 inches. Overall length was chopped from 213.7 to 209.9 inches, not quite two inches longer than the popular 1959 models. Weights dropped about fifty pounds.

Getting smaller wasn't unique to the 1961 market, for General Motors put its full-sized cars into the downsizing machine with results even more drastic. Chrysler Corporation was to do the same thing to its low-priced cars for 1962. However, the marketplace wasn't ready for such things at the time and smaller big cars were not sales successes, so as the decade progressed, so did the lengths and in some cases the widths of the full-sized cars.

Cars weren't the only thing smaller in 1961, so were new car sales. Another recession hit and model year production of domestic cars dropped to 5,407,256, their lowest level since the dismal 4,222,781 of 1958.

While production was down about ten percent from 1960 model year levels, the big Fords suffered even more erosion of their market share, with 791,498 being made, about thirteen percent lower than the previous year.

Ford's first high-performance volley was fired down the production line in the form of the 375-hp engine, the second came through the parts department.

Chevrolet and Pontiac had been offering multiple carburetion since the mid-1950's, Plymouth and Dodge went one better, putting a pair of four-barrels on ram-induction intake manifolds while Ford still had only a single quad to breathe through.

That situation was taken care of with the late 1960 announcement of the 6V option for the 390, just in time to be legal for the Winternationals. It consisted of an aluminum intake manifold with three Holley two-barrel carbs and a progressive linkage. It was the first such setup on a production Ford.

The center carb, which worked all the time, had a choke and was rated at 240 cubic feet per minute (cfm). The end units cut in about two thirds of the way open on the primary and had 300-cfm ratings, bringing the total to 840 cfm, well above the 600 for the single four-barrel.

Topping off the setup was an air cleaner for all three carbs, with an aluminum plate top and bottom and a minimal paper air-cleaner element sandwiched between. Advertised output was 401 hp at 6000 rpm with torque at 430 pounds-feet at 3500.

It was a bolt-on item and all else underneath remained the same. It also fit the earlier 332 and 352 engines, but in stock drag racing, would only be legal on the 1961 models.

Ford wasn't alone in its deadline-timed late announcement for performance parts. Chevrolet unleashed a 409-ci enlargement of its 348 and a new Super Sport option for the Impala; Plymouth and Dodge got the Chrysler 413 as an option with ram-induction ratings up to 375 hp; and Pontiac released a couple of hot dealer-installed options, a 368-hp 389 and a new 421-ci V-8.

More traditional styling marked the 1961 Fords, which had smaller external dimensions than the 1960 models. The sleekest of the bunch was this 1961 Galaxie 500 Starliner hardtop.

The hottest Chevy was rated at 409 hp, so after its seemingly big leap forward in cubic inches and horsepower only a few months before, Ford was in trouble again.

The first major confrontation between the Chevy 409's and tri-carbed Fords came in February at the Los Angeles County Fairgrounds in Pomona. With the largest drag racing crowd in West Coast history on hand, super stock competition turned out to be a battle between the two, leaving everything else, including Pontiac, by the wayside.

The Fords came close, but the Chevrolets took home the trophies. Frank Sanders of Phoenix, copped SS honors with a run of 13.63 seconds and trap speed of 105.26 mph. When they got into the elimination rounds, another Chevy driver, "Dyno" Don Nicholson, who would later play a major role in Ford's drag racing history, shut 'em all down in his 409 Super Sport.

To show how competitive the Fords were, Les Ritchey of Covina, California, turned an early run of 13.33/105.50 in his 401-horse Starliner.

Meanwhile, the stock-car types were eager to wring out the new 390. A Daytona test was arranged in the fall, again with The Wood Brothers preparing a new 1961 Starliner. Curtis Turner was hired to drive, and drive he did! He turned a lap of 153.505 mph, which was more than Jack Smith's one-lap record of 152.129. Turner then went 160 laps (400 miles) at an average speed of more than 150 mph. Needless to say, joy again reigned in the Ford camp.

Probably the best thing to happen to Ford's racing and performance program at the time had nothing to do with the cars themselves. An aggressive thirty-six-year-old, Lee Iacocca, took over as general manager of Ford Division. He felt Ford had lost its image as a performance car, knew of the boom coming in the "youth market" and felt racing was the way to get back on track in a hurry.

Ford spent the 1960 season building up an inventory of heavy-duty parts and extracting more horsepower from its engines. However, it left the racing up to the teams that chose to run its products.

Not having NASCAR teams under contract almost resulted in Holman & Moody running Dodges in 1961, despite the fact they built and sold Fords and had a warehouse full of parts.

Holman & Moody did have a contract with Autolite (which was before Ford bought it), and Autolite wanted Holman & Moody to race Dodges, for which Autolite was the main sparkplug supplier.

Iacocca realized Ford would have to get involved and, as would be the case throughout his career, he moved fast. He agreed to back Holman & Moody for the 1961 season. They hired Lorenzen, who used their garages as an independent in 1960. Lorenzen was out of money, sold his equipment and was ready to give up when he got the call. Holman & Moody liked his racing and mechanical ability and felt he could do the job.

It was too late to get a car ready for the 1961 Daytona 500, so the pits were scouted for a Ford for Lorenzen to drive. The search ended at the number eighty 1961 Starliner of Tubby Gonzales of Houston, Texas. It seemed that Tubby prepared a good car, but wasn't a front-running driver. After being promised a new Ford station wagon by Providence, Rhode Island, Ford dealer Bob Tasca, a performance-oriented Ford man if there ever was one. Tubby turned over the wheel to Lorenzen.

The patchwork effort wasn't enough to offset the Pontiacs, however. The Fords were the fastest ever, but the Pontiacs, with their smaller 1961 bodies, still had the edge. Fireball Roberts put a Smokey Yunick Pontiac on the pole at a record 155.709 with fellow Indian driver Joe Weatherly right behind at 154.122. Each took his respective twenty-five-mile practice race, each was followed by 1960 Pontiacs, which were still very fast. The fastest Ford in qualifications was Turner in The Wood Brothers Starliner at 153.4.

On the Friday before the 500, February 24, two hundred-mile races were held to determine the starting positions for the 500. The results were the same as the earlier events with Roberts taking the first and Weatherly the second, the latter in a record speed of 152.671 mph.

On Sunday the twenty-sixth, fifty-eight cars were ready for the 500 Grand National. Ford was again the most popular with twenty-one, followed by nineteen Pontiacs and fourteen Chevrolets.

Once the green flag dropped, Roberts out-classed the field in his 1961 Pontiac. Fords were able to run with the leaders, but various engine problems caused strong Ford pilots Turner and MARC convert Nelson Stacy to drop out early.

While Roberts couldn't be caught, Ford did have a chance for a respectable second, thanks to Banjo Matthews. But with eighteen laps to go, Matthews's engine let go and he was through. Five laps later the same thing happened to Roberts. Marvin Panch inherited the lead and took the win, driving Roberts's car from the year before, a 1960 Pontiac. It marked the second year in a row a year-old car won the race.

Panch averaged 149.601 mph, a record. Weatherly was second and Paul Goldsmith took third, both in 1961 Pontiacs. Lorenzen broke the string, bringing Tubby's car in fourth. Chevrolets did even worse, with a seventh the best finish. The 409's may have creamed 'em at Pomona, but they were also-rans at Daytona.

After Daytona, Holman & Moody got financial backing from Ford, as did The Wood Brothers, however the latter lost a driver when Turner was "banned for life" by NASCAR President Bill France after trying to organize the drivers to be part of the Teamsters Union.

The Ford effort in NASCAR was still not full-blown, as Lorenzen only entered fifteen of the fifty-two races on the schedule, but he won three, including two superspeedway events, the May 6 Rebel 300 convertible race at Darlington and the July 9 Atlanta 250-miler.

Fords only accounted for seven wins in 1961, but they took three of eight superspeedway events. The third was a surprising performance by Stacy in the Labor Day Southern 500 at Darlington. Stacy was running as an independent, but the win, in record time, helped get him a factory ride the next year.

The star of the Grand National circuit in 1961 was Pontiac, with its drivers in a record thirty events. Chevrolet was second with eleven. One of them was scored by Ned Jarrett of Newton, North Carolina, who switched from Ford to Chevy for 1961 and won the driving championship.

Pontiac's domination was also felt in USAC. Paul Goldsmith of St. Clair Shores, Michigan, drove Ray Nichels cars to ten wins in nineteen starts to take the driving title. In all, Pontiacs won fourteen of the twenty-two events on the schedule, leaving seven for Ford and one for Chevrolet.

Defending champion Norm Nelson left the Zecol operation to field his own cars and finished second in points with a pair of wins, aboard a Fairlane two-door sedan. He had some factory backing.

Zecol, which now was getting help from Ford, hired Don White and Tony Bettenhausen to drive for 1961. Bettenhausen was killed at Indianapolis practicing for the 500 and Eddie Sachs was signed. White drove a Starliner and scored three wins, Sachs usually was aboard a Fairlane and took two flags. Other drivers also took turns in the Zecol cars.

An event sanctioned by USAC, but at the time not counted in the point standings for the stock car division was the Pike's Peak Hill Climb, a twelve-mile race against the clock up the mountain.

Three of the first five stock cars were Fords, but unfortunately, it was a Chevy that won. The old master Louis Unser made it to the top in 15:06.0 in a 409-equipped 1961 Bel Air hardtop. Second was Turner in a Fairlane at 15:08.4. Then came a Chevy and two Fords.

The Fords were of national interest because they had a new option announced a few days before, the Borg-Warner T-10 four-speed manual transmission. They were among the first Fords so equipped. The option was dealer-installed and announced in time to qualify for the NHRA Nationals on Labor Day.

The unit was similar to that being installed on Chevrolets, Pontiacs and Studebaker Hawks, but there were differences. The tailshaft and housing on the Ford units were longer than the others and the gear ratios were different.

When equipped with the High Performance 390, 1961 Fords were popular competitors on the drag strips, but they had to run against some tough machinery. In later years they were downgraded from super stock to the lower classes. In the late 1960's, this 1961 Starliner owned by Laverne Schumann of Waterloo, Iowa, ran in H-Stock.

Ford's ratios were 2.37, 1.78, 1.31 and 1:1. They fell between Chevrolet's 409 gearing of 2.20, 1.66, 1.31 and 1:1 and Pontiac's 2.54, 1.92, 1.51 and 1:1. The difference was reportedly because GM was upset that Ford would be buying a unit from Borg-Warner that GM engineers helped develop.

Since stock cars on the big ovals run in top gear, the new four-speed would have little effect. It was intended to do some good in drag racing, where going through the gears is the only way to get to the end of the quarter mile.

Ford's new four-speed was aimed for the NHRA Nationals, which were moved to Indianapolis Raceway Park, in the suburb of Clermont. There were more Fords than ever in the top stock classes, but the results were surprising.

During the year a new class was added for the hot stocks, optional super stock racing, for non-production combinations. Hayden Proffitt took the class win in his lightweight 1961 Pontiac, but in the automatic O/SS class, Don Turner of Fortville, Indiana, won in a 1961 Ford, which was interesting because the hot 390's didn't come that way, as the automatics were believed to be too weak.

During the 1961 run, Pontiacs for both NASCAR and the drags were coming from the factory with aluminum sheetmetal up front and aluminum bumpers. Being smaller for the year with a 119-inch wheelbase, Pontiacs were weighing in lower than Fords and Chevrolets. This lightweight trend would continue for the next couple of model years.

Speaking of model years, it was time to change again, as the 1962 campaign was just around the corner.

1962

Ford started the 1962 run with an engine lineup similar to the way it ended the 1961 contest, except that the 401-hp option was now a production line package, as was the four-speed transmission.

With its intermediate Fairlane line being introduced for 1962, the lower end of the big Ford model lineup was trimmed. All sedans and hardtops now got formal roof styling and the Galaxie name.

At the base was the Galaxie series, which also was referred to as the Galaxie 100. It contained a two- and four-door sedan. Top of the line at the start was the Galaxie 500, which came in two- and four-door sedans and two- and four-door hardtops, plus a convertible. Gone was the sleek Starliner. As usual there was the full complement of the popular Ford station wagons, though the two-door wagon was omitted for 1962.

Styling was simplified and, in the opinion of the author, was the best of the decade for big Fords. Front sheetmetal from the 1961's was retained and the back

After meandering around for a few years, Mercury got back on track mid-year in 1962 with its S-55 series with bucket-type seats and console. It also got Ford's 406 V-8 as a mid-year option. The interior of an S-55 convertible had a sporty flair.

The 1962 USAC stock season boiled down to a race between Fords and Pontiacs. Each won ten races that year.

was smoothed-out with the fins being eliminated. A new grille filled the front cavity and rear styling continued the round taillight theme.

Base Galaxies had a chrome strip wrapped around the leading edge of the hood and continued within a few inches of the end of the rear fender. Galaxie 500 models had the same treatment, but top and bottom chrome strips were added above the front wheel and continued to the back, providing an area for contrasting two-tone paint. Galaxie 500's also got rocker panel and wheel lip moldings and additional roof trim. All sedans got bright metal window frames.

While there was no lack of chrome on the outside of the 1962 Fords, it was tastefully done. One part of the chrome that showed what was happening under the hood was an emblem for the cars equipped with 390 engines including the 300-hp Thunderbird 390 Special V-8, 375-hp Thunderbird 390 High-Performance V-8 and 401-hp Thunderbird 390 Super High-Performance V-8.

Those cars got emblems behind the front wheel well of the Thunderbird symbol with a pair of crossed flags and the numerals 390 in the center. The Bird part was done in red and black with black-and-white checkered flags. The unit was in silver chrome.

Competition for sportiness and performance was fierce during the 1962 model year, especially among low- and middle-priced full-sized cars. You could even get individual semi buckets in a Rambler.

Power-wise the Chevy 409 was now available in sufficient numbers after an unexpectedly limited run in 1961. Pontiac had so many versions and options of its big engines nobody could figure them out and even Buick released its 401-cid engine for its low-priced LeSabre.

While looking at the competition for 1962, it's worth noting that not everyone was shooting for the moon in horsepower and cubic inches. The standard Plymouth and Dodge Dart were chopped down in size, weight and engines, with the 361-cid initially listed as the largest available. Chrysler spent many dollars trying to convince the public that you really didn't need a big engine for good performance if the car was of sensible size and weight.

The concept had merit, but the styling didn't and the cars bombed badly for Chrysler. However, it would be these cars that would eventually revolutionize the high-performance motoring picture in the United States. Not because you didn't need a big engine but rather because if you put a big engine in a small car it would be faster than a heavier, larger car with a big engine.

Getting back to Ford, everyone who cared about such things knew that the initial 1962-model power situation would be a temporary measure until it came time to fire the big guns in December.

On schedule came the 406, which added 16 ci via a 0.08-inch bore increase. Two versions were available: the Thunderbird 406 High-Performance V-8 which was aimed at stock car racing and carried a horsepower rating of 385 at 5800 rpm and torque numbers of 444 at 3400. It had a single four-barrel carb.

For drag racers and brave street runners there was the Thunderbird 406 Super High-Performance V-8, rated at 405 at 5800 and 448 at 3500. Naturally it carried a trio of two-barrels.

Note that the tri-carbed unit, which would be classified for drag competition by its horsepower, only gained 4 hp over the hot 390, while the four-barrel unit was up ten horses.

Clean lines marked the 1962 Fords. While this Galaxie 500 two-door hardtop may have been good looking to the eyes, it wasn't to the wind, which made it slower than the other cars being raced that year.

The new engines replaced the high-performance versions of the 390. Blocks for the 406 were different with thicker walls. Compression ratio jumped from 10.6 to 10.9:1 with combustion chamber and bore changes. Pistons and connecting rods were tougher, the latter to take care of a problem that showed up in the NASCAR engines in 500-mile races. The intake valves remained 2.030 inches, but the exhaust valves grew from 1.560 to 1.625 inches.

Price for the 405-horse 406 and all the heavy-duty equipment which came with it (similar to that on the 1961's) was $379.70 over the 292 V-8. With a four-barrel on top, the tag was around $58 less.

One new option for 406 cars was quicker ratio steering. The standard manual unit came with a 30:1 ratio, but a 22:1 setup was optional. As before, no power steering was available. While quicker steering was desirable, it was mainly for circle-track racing applications and street drivers who opted for the lower ratio found themselves with an instant muscle-building device.

The media was waiting for the 406 and burned much rubber making their tests. *Hot Rod* staffers had the heaviest feet and did 0-60 in 6.5 seconds and the quarter in fifteen seconds flat with a 95-mph trap speed. *Car Life* hit sixty in seven flat and had a quarter run of 15.3/93, while *Motor Trend* was a hair behind at 7.1, 15.6 and 92.

The 406 answered the need for more performance (at least for the time being), but there was still the "bucket brigade" to contend with—the slew of models coming out with semi-bucket front seats.

Ford was ready for that too with its "Lively Ones," a group of mid-year models introduced in February. Falcon got the Sports Futura; Fairlane the Sports Coupe and a 260-cube V-8; and the big Ford got the Galaxie 500/XL series with two-door hardtop and convertible models.

The Lively Ones theme was from a variety television show Ford sponsored. While not terribly catchy, it did serve to call attention that Ford was combining its performance program with sporty models the average motorist could buy.

On the outside, the 500/XL didn't look all that different. A 500/XL emblem was on the rear gas tank filler door, and smaller ones replaced the "500" on the regular models, but there were big changes inside.

Ford finally put bucket-type seats in its full-sized car with soft vinyl, deep bucket full-foam cushioning. Chrome mylar piping was used in strips on the seatbacks and in door trim. A console was standard between the seats with a chrome ribbed panel up front and vinyl armrest/storage compartment between the seats. Rear seats carried the bucket theme, but were of bench design.

Doors had twin lights built into them with a white lens for illuminating the ground and a red one for oncoming cars. The dashboard carried the 500/XL logo on the glove compartment and pedals were trimmed in chrome.

Standard on XL's was the 292 V-8 and Dual-Range Cruise-O-Matic. Options ranged up to the 406's. Cars equipped with 406's got a gold version of the 390 plaque, done in gold and black with black-and-white flags.

Despite the late start, XL's proved fairly popular with 28,412 hardtops being built and 13,183 convertibles.

However, the decline in total sales for big Fords continued with 704,775 being built, down from 791,498 of the 1961 models. But Ford wasn't all that upset about it, for its launch of the Fairlane generated 297,116 orders, and most of them did not come out of the Ford tally, as had been the case with Falcon two years earlier.

These 1962 Ford convertibles figured in one of the major controversies of the 1962 season. The Holman & Moody cars of Nelson Stacy (#29) and Fred Lorenzen (#28) are parked after the May 12 Rebel 300 convertible race at Darlington, which Stacy won. They were later run in the Atlanta 500 with Starlift roofs which resembled the 1961 Starliner. Fred Lorenzen won the race and NASCAR declared the roof illegal for future competition.

So, 1962 may have been a good year for Ford in the showrooms, but it was anything but good on the racetracks and drag strips.

NASCAR banned Pontiac's aluminum body panels and bumpers for 1962, but that wouldn't help Ford very much—it had problems with its own cars rather than with the competition.

When the Starliner was dropped, so were Ford's superspeedway chances. The squared-off sedan roofs just didn't cut their way through the air as quickly as the sleek Starliners.

Pontiac also lost its slope-backed roof, but it didn't seem to make a difference. Chevrolet also had a formal roof, like the Pontiac, on its Impala hardtops, but left the 1961-style slanted hardtop roof on its Bel Air models.

When the gang got to Daytona it was the same old story, the Pontiacs were faster. They were running the 421's now and reportedly were getting 465 hp from them.

Chrysler gave in and released 413's again for its downsized cars and Chevy had 409 inches, leaving you-know-who at the bottom.

Ford's guns for 1962 were Lorenzen and Stacy driving for Holman & Moody, Panch with The Wood Brothers and limited help for independent Larry Frank.

Grand National activities at Daytona turned out to be a Fireball Roberts benefit. Still driving for Yunick, he set a qualifying record of 158,744 mph, won a hundred-mile preliminary in record time, averaging 156.999 mph, then blew away the 500-mile record and the rest of the field, averaging 152.529 mph.

Richard Petty finished second in the smaller Plymouth, followed by the new Pontiacs of Joe Weatherly and Jack Smith.

Ford drivers figured the Galaxie roof cost them about three miles an hour, but it turned out that they had more severe problems.

The bottom end of the 406 was not up to the task. Starting very early in the race, the hot Fords began dropping out with cracked blocks. Of the strong contenders, only Lorenzen was able to finish. He placed fifth, but not because he had a better engine—he cracked a fuel line and wasn't able to get his motor up to full revs.

As a clear demonstration of the relationship between racing and production at the time, Ford came up with a stronger bottom end for the 406. Main bearing web design on the center three mains was redesigned and four-bolt bearing caps were used. Up to then, two-bolt caps were utilized. The caps were cross-bolted, with the second set of bolts being attached to the skirt of the block. The bolts went right through and could be seen on the outside of the skirt.

Not only did the race cars get the four-bolt mains, but the change was made on the production line as well and late 406's installed in cars and sold over the counter also were cross-bolted.

The competition four-bolt 406's appeared by May and the Fords started finishing and winning. Although in all, Fords only won six Grand Nationals in 1962, four of them were major wins on superspeedways. Stacy took the May 12 Rebel 300 at Darlington (the last time it was a convertible race) and the May 27 World 600 at Charlotte. Lorenzen claimed the rain-shortened June 10 Atlanta 500 and independent Larry Frank notched the Labor Day Southern 500 at Darlington.

The star of the early 1962 model year was the Galaxie 500 Sunliner, shown here with a 390 V-8. Mid-year a bucket-seat-equipped XL series was added and a new engine came on stream, the 406.

Of all four wins, the most interesting was Lorenzen's Atlanta 500 victory, which demonstrated how far Ford would go at the time to win a race. The third annual running of the Atlanta 500 was originally scheduled for March 25, but it rained that day, as it did on the rain date, April 2. June 10 was the third choice, which fell after the May 12 Rebel 300 convertible race.

Not only were the 1962 Fords having engine problems, but the squared-off Galaxie roofs hurt the aerodynamics as well. Ford came up with a bolt-on roof for the convertible, the Starlift, with lines similar to the 1960-61 Starliner. It gave the option a part number and tried to get it legalized for NASCAR, even though the convertible's rear quarter windows didn't even cover the open area.

Ford took publicity photos, printed literature and even put Starlift-equipped convertibles in showrooms near racetracks.

Ford tried to use Starlift-equipped convertibles in the World 600, but was told the forty-five-day waiting period from the time of its April announcement had not elapsed. Other cars with notchback roofs were found for the Holman & Moody team, which was to use the Starlifts. It didn't turn out so bad; as Stacy won and Lorenzen took third.

By June 10, the waiting period ended and Holman & Moody brought the Starlifts to Atlanta, much to the discontent of the competitors.

NASCAR officials didn't quite know what to do. First they said X-members from convertible frames had to be welded in, then they said they couldn't decide if the cars were legal, then they said that the cars could be run for this race only, because Holman & Moody didn't have any other cars to run.

While all this foolishness was going on, the cars were unable to practice. When the race started, Stacy dropped out early with tire and chassis problems, but Lorenzen ran well and was ahead when rain stopped the event after 328 miles. Lorenzen got the win, but it was scored in a car declared "illegal" by NASCAR.

The Starlifts were parked for the year as far as NASCAR was concerned, but they would make one final appearance in the motorsports world.

On the day after the Atlanta race a newsman cornered Henry Ford II regarding the company's participation in racing despite the AMA ban on it. Ford admitted that his company was indeed involved and shortly after, the following statement was released by the public relations department:

> "The so-called 'safety resolution' adopted by the Automobile Manufacturers Association in 1957 has come in for considerable discussion in the last couple of years. I have a statement to make on this subject.
>
> "I want to make it plain that I am speaking in this instance only for the Ford Motor Company. I am not speaking for the AMA, of which I am currently president, or for the other manufacturers.

Accessories abounded for Fords in 1962. A few of them shown here include parking brake signal, power front seats, console range radio, power brakes and electric clock.

The winningest Ford driver in USAC stocks in 1962 was Norm Nelson who wheeled his own 1962 Galaxie two-door sedan to four flags. The notchback design wasn't as big a drawback on the shorter USAC tracks.

"Following the adoption of the AMA resolution, we at Ford inaugurated a policy of adhering to the spirit and letter of the recommendations contained in the resolution. We tried very hard to live with this policy. We discontinued activities that we felt might be considered contrary to the principles embodied in the resolution and also modified our advertising and promotion programs appropriately.

"For awhile, other member companies did the same. As time passed however, some car divisions, including our own, interpreted the resolution more and more freely, with the result that increasing emphasis was placed on speed, horsepower and racing.

"As a result, Ford Motor Company feels that the resolution has come to have neither purpose nor effect. Accordingly, we have notified the board of directors of the Automobile Manufacturers Association that we feel we can better establish our own standards of conduct with respect to the manner in which the performance of our vehicles is to be promoted and advertised.

"This action in no way represents a change in our attitude toward highway safety. Indeed, I think everyone is aware that Ford has been a pioneer in the promotion of automobile safety. We will continue, with unabated vigor, our efforts to design, engineer and build safety into our products and to promote their safe use."

Shortly after the announcement from Ford, Chrysler made a similar statement, leaving General Motors as the only one of the Big Three not to admit what had been going on.

The full effect of Ford's open involvement in stock car racing wouldn't be apparent until the 1963 season. When the final 1962 NASCAR Grand National totals were in, Pontiac was again king with twenty-two wins, followed by Chevrolet with fourteen and Plymouth with eleven. Pontiac's Joe Weatherly won the driving championship.

Ford's output of six wins was its worst since the factory first entered NASCAR racing in 1955 and two victories were scored.

Things weren't quite so bad in USAC stocks, where the aerodynamics of the Galaxies on the shorter tracks weren't a factor. Paul Goldsmith repeated as champion in a Nichels Pontiac, taking eight wins. A. J. Foyt and Rodger Ward also won events in Pontiacs.

Ford, however, also accounted for ten wins to tie Pontiac. Don White, driving for Zecol, gave Goldsmith a run for it in the point race and took two wins on the way to a second-place finish. Norm Nelson, still driving his own car, notched four wins for Ford and placed third. Bill Cheesbourg also took two wins for Ford, with single flags going to Troy Ruttman and Foyt.

Separate owner standings were kept in 1962 and Zecol won that title for White's Galaxie, as Goldsmith drove more than one car.

Another honor that went to Ford in USAC was the Pike's Peak Hill Climb. Curtis Turner's time of 14:55.5 topped the competition in the stock class and a Ford 406-powered sports car driven by Ford performance advisor Ak Miller took the over-two-liter sports car class.

Mid-year the 1962 Galaxie 500/XL joined the full-sized Ford lineup. Inside there were bucket-type seats and a console, outside identification was minimal. This prototype does not have the correct roof emblem and looks like it doesn't even have seats.

In drag racing the rules changed, but the results for Ford didn't. In the NHRA Winternationals at Pomona February 17 and 18, the new Ford 406's were strong, but it would again be the Pontiacs and Chevrolets taking the big wins.

Two new types of classes greeted the drivers. Most popular was factory experimental. This involved combinations of engines and cars the factories didn't build. For example Hayden Proffitt of Westminster, California, won A/FX in Mickey Thompson's 1962 Pontiac Tempest coupe with a 434-cube V-8 stuffed under the hood. Don Nicholson took B/FX in a Chevy II wagon with a 360-horse 327 aboard.

As the season progressed, so did the technology battle. Lighter Chevrolets and MoPars were being drag raced and Ford reacted by having ten lightweight Galaxies built for the Labor Day Nationals.

The cars were constructed in Andy Hotton's Dearborn Steel Tubing shops and reportedly weighed 3,320 pounds. The Galaxie two-door sedans had fiberglass front fenders, hood, rear deck lid and aluminum inner fender panels and bumpers. Fiberglass doors were planned, but weren't ready in time. The engine also had special equipment including an aluminum intake manifold with two four-barrels.

The lightweight Galaxies were thrown into A/FX with the Tempests and were eaten alive.

Super super stock competition was the same old story. Fords didn't do badly in the early rounds, but lost out when it counted. Dave Strickler's Bel Air copped stick honors and Al Eckstrand pushed buttons on his way to an automatic victory in a 1962 Dodge Dart.

Proffitt was wheeling Bill Thomas Chevrolets by Labor Day and brought the Indy crowd to its feet in the final stock eliminator run-off by downing Eckstrand's Dodge in a close match.

Before discussing the 1963 models, a sideways glance at Mercury is in order. In the horsepower race days of the 1950's, Mercury was in there battling door-to-door with Ford. In fact there were factory-backed Mercurys racing in 1955 before the Fords took to the track.

In 1957, when the AMA cutback hit, factory Mercurys were being raced and sold with the M-335, a 368-cube V-8 with 335 hp. Announced for 1958 was the industry's first 400-hp engine, this one based on the 430-cube Lincoln unit.

Production of the Super Marauder never got off the ground and Mercury's involvement in performance ended. Styling themes varied in the coming years with the "Big M" cars lacking direction.

When Ford got back into the performance biz in 1960, Mercury did not. It wasn't until mid-year 1962 that Mercury was back on track. Up until then the Monterey 390 Police Special at 330 hp was as hot as she got, and then it was supposedly only for the law-enforcement types.

Shortly after Ford announced its 406, Mercury finally put together a performance program.

Both the 385-hp four-barrel 406 (Marauder V-8) and 405-hp (Marauder 406 V-8) were offered in all models except station wagons. Prices over the 292 V-8 were $321.80 and $406.70, respectively.

As in the Fords, you also got a heavy-duty chassis, bigger brakes and stronger suspension. The Borg-Warner T-10 was a mandatory option at $220.90. Incidentally, the price for the same gearbox in a Ford was $188.

A ploy of the Ford Motor Company to legalize a fastback roof for its convertible race cars and gain an advantage on superspeedways was the Starlift roof, which bolted onto the stock convertible. The lines were similar to the 1961 Starliner and were more aerodynamic than the 1962 notchback roofs. These photos showed a Starlift above and on a Galaxie 500 Sunliner. The option was not allowed by NASCAR and was only used in one race. It never was widely available, as the rear windows didn't even match. Sharp eyes will detect Thunderbird wheel covers.

Mercury also got a mid-year line of bucket-seat models for its full-sized Monterey, the S-55 in two-door hardtop and convertible models.

There was no Mercury racing program for 1962, but that would change before too long with the introduction of the 1963½ models.

Before the 1962 season ended, both NASCAR and NHRA announced that there would be a seven-liter (about 428-ci) limit on the 1963 stock cars. This was to stop some of the rumors about engines that were ready to go into production of up to 500 cubic inches.

With the size of its big engines no longer classified information, Ford made public a 483-cube monster it had built, based on the FE block.

It was installed in one of the Holman & Moody Starlifts and run at Ford's Romeo, Michigan, Proving Grounds at 182 mph. Ford figured it could break some records with the beast as it fit below the eight-liter (488-ci) limit for national and international Class B speed records.

On October 2, 1962, the Starlift was run at the Bonneville Salt Flats with Lorenzen, White and Ralph Moody sharing the driving chores and USAC doing the timing.

A total of about 500 miles were run before the engine blew and forty-six records were snapped along the way, some dating back to 1935 when Ab Jenkins set them in a Duesenberg Special. The average speed was 163.91 mph for the distance. A top speed of 176.978 was recorded by the team.

White had some experience with speed runs prior to the test, as he drove a brand new 1961 Starliner to a record 159.320 mph in the last Speed Trials at Daytona Beach in 1961.

The Starlift (carrying Lorenzen's number, twenty-eight) used 2.67:1 gears and 8.00x15 tires. The 483 had a 12:1 compression ratio with a bore and stroke of 4.23 and 4.3 inches. A pair of four-barrel Holleys were on top.

1963

The two-stage model year reached its peak for the full-sized Fords in 1963. Both for engines and body styles, the best was saved for mid-year introduction.

Starting things off was a lineup of facelifted 1963 Fords. All exterior sheetmetal below the beltline was changed, but the same basic body shell was used. The round taillights grew, the hood was flattened and the grille redesigned with yet another stamped pattern.

Model lineup was generally the same, but a four-door hardtop was added to the Galaxie 500/XL series, complete with buckets.

Chrome trim changed with the Galaxies getting a full-length strip down the side at door handle level that doubled back after getting to the taillight. Other trim changed as well. Base Galaxie sedans no longer came with standard chrome on the window frames. The 390 emblem was done in silver and black with checkered flags in black and white.

Other changes in the 1963 models included a new dashboard design, the option of a swing-away steering wheel and crank-operated front vent windows standard.

This 1963 Ford Galaxie 500/XL sports hardtop with the vinyl top option is a perfect example of the elegant and lavish ride found in Fords of this year.

Length was up 0.6 inch to 209.9 and width 0.8 to an even eighty inches.

Mechanically there were changes as well. Gone was the early Y-block 292. At first it was replaced by the Challenger 260-cid V-8, which produced a rating of 164 hp, but later in the model year the 195-hp 289 was used as the base V-8. The 260 started as the standard XL powerplant and the 289 finished in that role.

A new Ford all-synchromesh three-speed manual transmission was standard with all engines except the 406's, which started the model year as the top line offering and again only came with a four-speed bolted on behind.

The front suspension was redesigned with the lower front control arm having a crank-like mounting so it could give horizontally as well as vertically, a system used on the Thunderbird in previous years.

While magazine awards are often not worth noting because they usually are made to promote the magazine more than the car that gets them, one given to the 1963 Ford was interesting.

Auto Sports gave its First Annual Award of Excellence to the 1963 Ford Galaxie 500/XL two-door hardtop with the 406-cube, 405-hp engine. The test car did 0-60 in 7.9 seconds (which wasn't unusual for cars of the day) and ten pages were spent praising its leadership in the performance field.

Apparently the whole test, evaluation and award were done at the press preview in summer and without the knowledge that Ford's real ammunition was being saved for announcement later in the model year. For while the notchback 406 hardtop wasn't all that bad, it was pale in comparison to Ford's 1963½ offerings, which were announced late in the calendar year.

Up to then, Ford had been just warming up—it took the gloves off with the 1963½ models. Not forgotten were the aerodynamics of the Starliner nor the new racing-engine size limit.

Leader of the Galaxie line was the new Sports Hardtop, a two-door hardtop with sloping rear pillars and small formal-type rear window. At 54.5 inches high, it was an inch lower than the notchback hardtops. Though technically it was not that type of design, it was referred to as the fastback.

Available in both the Galaxie 500 and Galaxie 500/XL series, it was the answer to the needs of stock car drivers on both superspeedways and drag strips.

Also for that group, and the folks on the street with a heavy foot, was a new size of engine, 427 cubes, just below the newly announced sanctioning body limit. It was arrived at by increasing the bore of the 406 from 4.13 to 4.23 inches which, considering the block had 4.63-inch bore centers, was really pushing its bore capacity and narrowing the space between the cylinders for cooling.

As was usually the case, the 427 replaced its predecessor, the 406, which lasted only about a year in the rough high-performance ring.

Cars with 427's got plaques similar to the 406 in gold and black, but with the new numbers in the center.

For 1963 this emblem stood for the lusty Thunderbird 427 V-8!

Deluxe wheel covers for 1963 were offered in two styles: handsome, fully chromed simulated wire-wheel type (shown here); and the full-disc design.

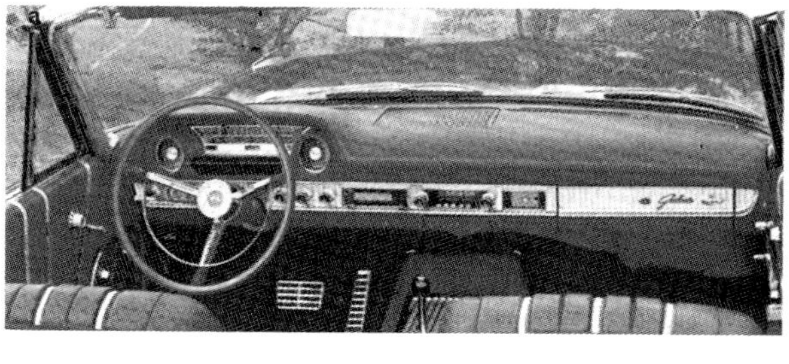

Ford Galaxie 500/XL instrument panel for 1963 combined style and function; instruments were arranged for quick reference, set in a deep cove and softly lighted to eliminate glare.

Initially, there were three versions of the 427 available, two for the production line cars and one for the lightweight drag racing machines.

Production 427's came in single four-barrel and new twin four-barrel versions. They shared the same block with cross-bolted mains, crankshaft, rods, pistons, 276-degree camshaft and 11.6:1 compression ratio. Also, both got a new exhaust manifold with better flow than the old units used on the 352, 390 and 406 performance versions.

To qualify for NASCAR, USAC and other stock car racing, the single four-barrel unit was available with a rating of 410 hp at 5600 rpm. Torque worked out to 470 pounds-feet at 3400.

The triple two-barrel setup served well as an interim intake manifold for Ford high-performance engines, but when the cubes got up to 427, more venturi area was needed and dual quads did the job in a less complicated manner. A pair of 540-cfm Holleys sat atop a new aluminum intake manifold. The rating was a figure that would soon be a common one in the industry, 425 hp; Ford's was attained at 6000 rpm. Torque grew to 480 at 3700.

Rarest of the three 427's was the drag version. Advertised output was the same as the dual-quad street engine (remember the drag rules), but the powerplants came with 12:1 compression, 600-cfm Holleys and a 300-degree camshaft. We'll look at the package they came in a little later.

With the 427 came an end, at least for awhile, to Ford's problems of always being a few inches behind in the displacement race. The racing limit put a lid on the cube competition and forced the manufacturers to look to the insides of their engines, instead of the boring bar, for added output.

Chrysler introduced a 426-ci version of its wedge engine for Plymouth and Dodge performance efforts, Chevrolet brought out a new and short-lived 427-cube version of the 409 and an even scarcer all-new Mark IV engine, before GM cut its performance programs in a drastic move. Though not much of a factor, Buick also introduced an engine under the limit at 425 ci. Mercury shared the Ford 427, so there was lots of company around the seven-liter mark.

In the regular Galaxies, checking the 427 box on the order form still got you a host of goodies, including 7.10x15 nylon tires, compared to 7.00 to 8.00x14 tires on the other models, a heavier frame, heavy-duty suspension, stronger brakes and a heftier driveline. With the 427 you still had to get a four-speed and you still were unable to have the big mill in a wagon.

Cost for the 410-hp setup was $405.70 over the standard V-8, while the 425-hp job ran an additional $461.60 over the base V-8. Four-speeds still listed at $188.

Ford's Equa-Lock limited-slip differential was not up to the 427's output, so buyers wanting the feature had to go the dealer-installed route and shell out $110 plus installation for a Dual-Drive setup from Detroit Automotive Products.

The dealer performance option list grew as Ford got deeper into the speed scene. One interesting option was exhaust cutouts that bypassed the muffler. Supposedly only for use at the strip, they listed for $55.

New sheetmetal graced the 1963 Ford line with large taillights dominating the tail section. A Galaxie 500/XL shows off the new lines.

Ad copy writers were as active as the engineers. The 1963 models started as "America's liveliest, most care-free cars," but with a concept of building and racing all types of cars from Monte Carlo Falcons to Indianapolis 500 powerplants, it wasn't nearly as descriptive as the "Total Performance" campaign launched later in 1963.

Just for the record, 1963 Galaxies, after their injection of 1963½ goodies, became the "Super Torque" Fords.

Not all the 1963 mid-year offerings were aimed at the moon. Ford also introduced a low-priced base series in its full-sized line shortly after the start of the model year, the 300. It consisted of stripped-down two- and four-door sedans, but was not the basis for any major performance projects.

As in 1962, Mercury paralleled Ford's mid-year offerings. Its early 1963 model sedans and hardtops featured "Breezeway Design," with reverse-slanted, opening back windows. All models were on a 120-inch wheelbase.

Introduced early in calendar 1963 was the fastback-type two-door hardtop for both the base Monterey and bucketed S-55 series. Also added were 427's replacing the 406's. The Mercury name givers were hung up on the Marauder name, giving it to both the new hardtop design and all the V-8 engines. The four-barrel 427 was billed as the Marauder 427 V-8 and the dual quad got a Marauder Super 427 V-8 label.

Seeing the lack of Ford success in drag racing, Mercury decided to send its Marauder Mauraders stock car racing, where its being a bit bigger and heavier than the Fords shouldn't have been as much of a problem.

Of course Ford was eager to go racing with its new hardtops too. Holman & Moody went to Daytona in November (ahead of the public introduction) to test a new 427 hardtop and toured the track at 161.3 mph, which was faster than the 160.7 track record set by a 1962 Pontiac.

In comparison, a 1962 Galaxie 406 went 155 and a 1963½ hardtop with a 406 was clocked at 159. Holman & Moody figured the airflow over the top of a 1963½ was twenty-three-percent cleaner than the 1962 Galaxie sedan.

The first major event of the 1963 season was a new one, the Motor Trend 500 on the 2.7-mile road course at Riverside (California) International Raceway on January 20. Both the Ford "fastbacks" and Mercury Marauders (prepared by Bill Stroppe) were there and stole the show.

New Pontiacs and MoPars were also on the scene, but took a back seat to the Ford products. Sports car ace Dan Gurney, who drove at Daytona for Holman & Moody in 1962, put a Ford on the pole with a record speed of 99.590 mph. He

With glistening chrome accessories, a 1963 Ford 427 nestles under the hood of a new Galaxie. Several Ford engines had dress-up kits available over the years.

Don White won the 1963 USAC stock car championship in a 1963½ Ford fastback. His Zecol-Lubaid ride had Ford factory backing.

then led 120 of the 185 laps for his first NASCAR win. Mercury's Parnelli Jones was the only driver to give him a run for it before he dropped out with transmission problems.

Foyt was second in a Pontiac and Troy Ruttman was a lap back in third in a Stroppe Mercury.

The next major event in racing for the 1963 season didn't take place on a racetrack but rather somewhere in the General Motors corporate headquarters.

Right after Riverside an edict came down to division general managers to get out of racing, pronto! This only applied to Chevrolet and Pontiac, but would affect racing and the entire performance car market in the immediate future.

Back-door assisting of racing teams also came to a halt; this time GM brass was serious, much more so than when the AMA ban took place in 1957.

Reason for the cutback had to be tempered with the knowledge that both Ford and Chrysler were going all-out in the performance and racing fields and while GM could keep up when everybody was dealing under the table, it would be impossible to do so now, unless GM came out and disavowed the AMA ban like its competitors.

It must also be remembered that GM was in constant fear of being broken up by antitrust action, as it had over half the new car market. So it tried to maintain a low profile where it could so officials in Washington wouldn't be attracted to take a closer look. Blowing Ford and Chrysler into the weeds in the booming sport of auto racing just didn't fit the GM game plan.

The hottest Pontiac and Chevrolet engines were axed as part of the change, but that didn't mean that the two makes wouldn't be heard from again, especially when it came to vehicles for the street.

That also didn't mean some of the hot engines didn't get out before the cutback took effect, for when the Fords got to Daytona with the greatest confidence of the decade so far, they were in for a real surprise.

Chevy drivers had new 427 Mark IV engines in their Impalas, porcupine-valve heads and all. While the Fords practiced at the predicted 161 mph, the Chevrolets were in the 164 range. Even more aggravating was the fact NASCAR let the engines run, despite the fact they were unavailable to the public. They were not production and not even available over the parts counter.

Junior Johnson put Ray Fox's Chevy on the pole with a record average of 163.681 mph, some 10 mph over the fastest 1962 Chevy. Even faster, but missing out on the pole-day timing was sprint car driver Johnny Rutherford, who turned 165.183 in Smokey Yunick's Chevy. Fastest of the Ford pilots was Fred Lorenzen at 161.870.

Despite the prerace controversy, fans quickly forgot the Chevrolets and concentrated on the Horatio Alger story of Tiny Lund. The regular pilot of The Wood Brothers Ford, Marvin Panch, was driving in a sports car preliminary practice session when his Maserati GT crashed and landed upside-down in flames. Several men watching nearby ran to the car, turned it over and got Panch out before the safety crews got there. Among them was DeWayne "Tiny" Lund who, at thirty years old, had driven in Grand National races the last eight seasons, but not with any success.

Ford's man in International Motor Contest Association (IMCA) competition was Dick Hutcherson, who won the stock car title in 1963 and 1964 before moving on to NASCAR. He is in his 1963 car, a 1963½ fastback.

Panch's injuries would keep him out of the 500, but not from returning to racing later in the season, thanks to the quick work of Lund and the others. From his hospital bed, Panch, who won the 500 in 1961, asked The Wood Brothers to give Tiny the ride for the 500.

With reservations, they put the 265-pound Lund in the car, taped over Panch's name and hoped for the best.

Probably no one expected the result: Lund won the race; his first Grand National victory and the biggest of his career. The Chevrolets dropped by the wayside with mechanical problems and Fords finished first through fifth.

Though Fords went on to dominate the rest of the Grand National season, that didn't mean there weren't some interesting moments.

Despite the GM cutback, there was some Chevrolet and Pontiac competition for the Fords. Johnson and Fox kept the Mark IV-powered Chevy going all season, entering the major races. Being an underdog against the factory Ford teams and usually the fastest car on the track made Johnson the hero of NASCAR. He won superspeedway races, the Dixie 400 at Atlanta and late season National 400 at Charlotte. After the last race the team was out of parts.

Pontiac stalwarts Joe Weatherly and Fireball Roberts didn't last the season. Weatherly won the Rebel 300 at Darlington, then brought his chief mechanic Bud Moore with him to a deal with Mercury. Moore would later play a key role in Ford's racing programs.

Roberts gave up his Indian for a ride with Holman & Moody, winning two superspeedway events: the July Fourth Firecracker 400 at Daytona (where a revised 427 Ford engine premiered) and Labor Day Southern 500 at Darlington.

Lorenzen was the star of the circuit, taking six wins including the Atlanta 500 and World 600 at Charlotte. He became the first stock car driver to top $100,000 in single-season earnings, collecting $113,570.

Fords won twenty-three races for their best season since 1957. Plymouth was next with nineteen, but had to earn them on the short tracks. Then came Chevrolet with eight, Pontiac with four and Mercury with one.

Mercury's foray into NASCAR had not gone well. Its only win came in the final event of the season, the Riverside Golden State 400 on November 3. Drivers aboard Mercurys included Darel Dieringer, Parnelli Jones, Rodger Ward, Rex White, Johnson (last race) and Weatherly. Dieringer won, but it was too late. The decision for 1964 racing had already been made to cut back the stocks and enter Comets in A/FX drag racing.

Weatherly won the driving title for the second straight year.

The USAC stock season was a mixed bag as far as makes of cars went. Don White drove a Zecol 1963 Ford to the driving championship, but only scored one win. In all, Fords won five races, topping USAC's performance index, but USAC, which always did things differently, gave its manufacturer's award to Plymouth (which finished second in the index and points total) because Plymouths won six races.

Mercury drivers won four races with Jones taking three of them, all in Milwaukee (Elmer Musgrave won the other). One race went to a Pontiac, which was driven by defending champ Paul Goldsmith.

With his name on tape on the door, DeWayne "Tiny" Lund poses next to the 1963 Ford Galaxie fastback he would drive to victory in the 1963 Daytona 500. Lund was named a substitute driver after he helped rescue the regular driver of The Wood Brothers #21, Marvin Panch, from a fire. Fords took the first five positions; Lorenzen was second in a Holman & Moody car.

Mercury entered stock car racing competition with factory-backed teams in 1963. Behind the wheel of a Bill Stroppe 1963 Marauder hardtop at Daytona is Parnelli Jones, who ran most of the year in USAC.

Jones topped the stocks at Pike's Peak in a Stroppe Mercury, giving that make its only claim to fame up to that point. After that, advertising for the big Mercurys carried the theme, "Pike's Peak Champion." Turner, who ran in USAC that season and finished fourth in points, was second in the hill climb in his 1963 Ford. Ak Miller topped the sports cars again in his Ford-powered Devin.

Ford's fortunes in drag racing in 1963 didn't improve much, but it wasn't that Ford didn't really try. It was just that a combination of left-over Pontiacs and Chevrolets and newly potent Plymouths and Dodges were too much for the big Galaxies.

The main ammunition would be a lightweight version of the Galaxie fastback hardtop with a special engine. It would qualify for NHRA super stock competition. NHRA simplified things a bit, dropping super super stock and going back to S/S as its top stock class. Nonproduction units would still run in the factory experimental division.

To do the job, a run of fifty fastback hardtops was made at Dearborn Steel Tubing. To start with, a lighter Ford 300 frame was used. Fiberglass doors, hood, trunk lid, front fenders and inner liners replaced the steel panels. Aluminum bumpers and brackets also cut weight. Thin-shell bucket seats were up front with a regular bench back seat. Thin rubber floor mats covered the floor and all sound-deadening material was left off the car.

Options like radio, heater, clock were omitted. The four-speed transmission got an aluminum case and stronger gears with its own ratios of 2.20, 1.66, 1.31 and 1:1. An aluminum bell housing was cast by R. C. Industries and weighed thirty-four pounds. Heavy-duty springs, shocks and fifteen-inch wheels were continued from the standard performance models. Early models lacked the fiberglass doors and inner fenders and came with cast-iron headers. They weighed in at 3,510 pounds. Later models were 3,425.

At first the cars ran in A/FX, but when production was completed, switched over to S/S competition.

Les Ritchey made early model runs at 12.29 seconds and 117.30 mph, but Gas (Gaspar) Ronda lowered that, when all the goodies arrived, to 12.07/118.04.

As usual, the first acid test for the new equipment was the Winternationals at Pomona and as usual, the new Fords came close, but did not win in the hottest stock classes.

Lightweight MoPars with 426's were proving very tough and nearly unbeatable in the top automatic classes.

Bill Shrewsberry topped A/FX at Pomona in a Pontiac, while both S/S class wins went to Plymouths with Tom Grove the best among the shifters and Bill Hanyon king of the automatics. In the stock eliminations, Al Eckstrand prevailed in a 1963 Dodge, knocking off a Plymouth in the final round. Eckstrand shook up the troops as his mount had an automatic and any good drag racer at that time firmly believed that there was no way an automatic could beat a properly shifted stick.

The ninth annual NHRA nationals at Indianapolis Raceway Park again saw

A new roof line with slanted rear pillars highlighted the 1963½ Ford Galaxie Sports Hardtop, shown in Galaxie 500/XL trim. Dubbed the "fastback" it was an immediate hit in show rooms and on the racetrack. In the latter case it solved an aerodynamic problem Ford had with the 1962 models.

the Fords shut out, but boy did one come close! In the S/S final, John Barker's Dodge shut down Ed Martin's Ford 427 to get the class win, but later Barker was disqualified for having an illegal cam. NHRA rules did not elevate the second place car to the win.

An interesting vehicle there was a lightweight version of the Fairlane entered by Bob Tasca, but that's a story for the Fairlane section.

One place the 1963 Fords weren't losers was in the showroom. Production for the model year ended at 845,292, up nearly twenty percent from the 1962 tally of 704,775. It was the first increase for big Fords since the 1959 models.

Almost all of the increase was accounted for by the fastbacks, which rolled up 134,370 orders (100,500 for the Galaxie 500 and 33,870 for the 500/XL). The Galaxie 500 total was the second highest in the full-sized Ford line, topped only by the Galaxie 500 four-door sedan.

1964

Total Performance marched on for the 1964 model year and with it the full-sized Fords, which got their usual quota of new sheetmetal. Two-door hardtops kept the 1963½ roofs, and notchback two-door hardtops were dropped. Four-door hardtops and all sedans got new roof designs with less squared-off lines. Below the beltline all sheetmetal and bumpers were new.

Yet another full-width grille was up front, this one with horizontal bars with three crinkles in it. Along the sides was a large concave area, highlighted by a convex spear that started with the headlights and tapered back to the center of the front doors. Various trim was used on the spear.

At the rear were the familiar large round taillights, located at either end of the large concaved section which ran the width of the car.

When cars had the 390 or 427, the Thunderbird insignia plaques were located on the lower edge of the front fenders, just in front of the doors.

Despite the visible alteration, external dimensions changed little. Width remained at eighty inches and length decreased 0.1 inch to 209.8.

The model lineup was shuffled on the bottom end. Two- and four-door sedans were in both the base Custom and mid-range Custom 500 series. The Galaxie 500 continued with two- and four-door sedans and hardtops, plus a convertible. The three Galaxie 500/XL models available were the convertible and two hardtops (two- and four-door).

Dashboards received minor changes, but the bucket seats in XL's were redesigned with thinner shells and less padding. A console, 289-cid V-8 and Cruise-O-Matic were standard on the XL's. The two-speed Fordomatic previously standard was no longer available in the 1964 full-sized Fords.

Engine-wise, the production big Fords were little changed from the 1963 versions. The 352 picked up a four-barrel carb and compression increased from 8.9 to 9.3:1 for a new rating of 250 hp, thirty over the 1963 tally.

With exterior highlights like spinner wheel covers, fairly reserved chrome trim and styling, the 1964 Galaxie 500/XL convertible was an attractive package.

The 390 still carried a 300-hp rating, marking the seventh consecutive year an engine in the Ford line was given that number. Police departments were still able to buy the 330-horse Interceptor version of the 390.

For once, production line big Fords didn't have a big happening at mid-year. But that didn't mean the racing versions didn't undergo a shaking up.

Tests of the 1964 427 street Galaxies didn't reveal much progress. *Speed & Custom* put one through the paces, a Galaxie 500 two-door hardtop with a dealer-installed Detroit Locker differential and 3.25:1 gears. The testers were disappointed with a 0-60 run of 7.1 seconds and quarter-mile performance of 13.96 seconds and 106.8 mph.

The drag racing engine underwent more changes than the street version. New high-rise heads and intake manifolds, plus a new cam helped real output. A larger Holley carb was used, intake passages were enlarged and machining was done around the valves.

The trouble with the drag racing 427 High Riser was that with the air cleaner on, it didn't fit under the hood. That kept it out of production cars. However, on the specially built lightweight Galaxies, Fairlanes and Falcons that would be built that year, all had fiberglass hoods with big bubbles for proper clearance.

Another new feature of the lightweights was the conversion of the inner headlights to air intakes with ducts leading from them to either side of the aluminum air intake over the twin four-barrels.

NASCAR stockers got modified high-rise intake manifolds for the four-barrels, and were able to retain the stock hoods.

Mercury restyled the front and rear ends of its full-sized cars. It changed the model lineup, going back to the old Monterey, Montclair, Park Lane nomenclature it used in the 1950's. Gone was the S-55, but the top-line Park Lane could be had with thin-shell buckets similar to those in the Ford XL.

Four-door hardtops came in both Breezeway and fastback styling. The fastback hardtops gave yet another opportunity to attach the Marauder name to something. With fastback two- and four-door hardtops in all three series and six different Marauder engines, there were now a dozen Marauders of one kind or another available.

Mercury continued its stock car program for the big cars, but placed most of its performance emphasis on the compact Comet, which would be used for endurance runs and drag racing.

Both Ford and Mercury were big cars; however, much of the performance news for 1964 would be made in smaller cars, a trend that would eventually take the big ones out of the spotlight.

Chrysler, which knew it had a good thing with its smaller cars and big engines, started off the 1964 model year by offering 365-hp street versions of its 426 wedge in its Plymouth and standard Dodge lines. For the racers it sold the 426 III, which was an improvement of the 426 II mid-1963 engine. And that was just a warmup for its biggest bomb of all, the 426 Hemi, which came out mid-year.

General Motors brought out all-new intermediates with 115-inch wheelbases: the Chevrolet Chevelle, Pontiac Tempest, Oldsmobile F-85 and Buick Special. Though the first ones came with engines no larger than 330 ci, there was room under the hood for just about anything GM made.

Pontiac didn't wait long to prove it; shortly after the start of the model year, it brought out the GTO option for the LeMans series. The main feature was Pontiac's 389-cid V-8 with ratings up to 348 hp.

Mike Schmitt captured the 1964 NHRA Junior Stock World Championship in his 1964 Ford Galaxie 500 two-door hardtop with a 427 under the bubble hood. He pulls up to the line at a late 1964 meet. Schmitt successfully campaigned Galaxies for several seasons.

Ned Jarrett turned out to be Ford's winningest NASCAR driver for 1964, wheeling his Bondy Long Galaxie to fifteen victories, half of Ford's circuit-leading total of 30 for the year.

There was no factory racing program for the GTO, but the street buyers didn't seem to care. Sales took off, imitators followed and soon the same thing was happening on the street as on the drag strip, smaller cars outdoing the bigger ones.

Performance buyers make up a relatively small minority of the buying public, so there was no downturn in big Ford sales. Indeed, production was up 9.2 percent to 923,232. The new car production rate for all domestic 1964 models was only up 7.5 percent.

A bigger surprise was that for the first time, the best-selling model in the full-sized Ford lineup was a two-door hardtop. The Galaxie 500 fastback produced 206,998 orders to 198,805 for the runner-up Galaxie 500 four-door sedan.

No doubt contributing to the popularity of the 1964 Galaxie two-door hardtops was their use in stock car racing. With rapidly growing interest in racing and Ford's heavy promotion of its stock car accomplishments, the hardtops took on the image of a winner. A buyer may have ended up with a 289 under the hood, but at least he had a car just like Fred Lorenzen, Fireball Roberts or Don White drove.

The first major win for the 1964 Ford was the Motor Trend 500 at Riverside. Dan Gurney switched over to a Wood Brothers Galaxie, but the result was the same and he scored his second straight win. Teammate Marvin Panch finished second, giving the Woods a one-two finish.

Unfortunately, Gurney's win was overshadowed by the tragic loss of Joe Weatherly, who hit a wall in his Bud Moore Mercury and was killed. Moore continued his factory-backed efforts in 1964 with his number two driver, Billy Wade.

For the second year in a row, the Ford teams confidently pulled into Daytona, only to be confronted with another fast engine that was not available to the public.

Darel Dieringer had a factory Mercury ride and won a single NASCAR event in 1964. He also drove 1964 Mercs in 1965 after the factory stopped backing the cars.

1964 Ford Galaxie 500/XL convertible—fine car luxury plus all the most-wanted features made this the stand-out convertible in its field!

Mercury stock car racing plans for 1964 included Parnelli Jones driving for Bill Stroppe in USAC and Joe Weatherly in a Bud Moore car in NASCAR. Jones (right) is shown at the wheel of his Marauder, which he took to the USAC title. Weatherly unfortunately was killed in the opening race of 1964 at Riverside, California.

Plymouths and Dodges had 426-ci Hemis stuffed under the hood, which produced considerably more horsepower than the Fords.

Back in the 1950's, the most powerful engines of all were the Chrysler FirePower V-8's, which had hemispherical heads. When the horsepower race cooled and production costs were analyzed, it was cheaper to produce Chrysler's new line of wedge big blocks. The FirePower was dropped after the 1958 model year.

While Detroit may have forgotten about the Hemi, the drag racers hadn't. The deep-breathing heads with twin rocker shafts produced lots of horsepower and continued to be used by a large number of competitors, especially the top fuel dragsters.

When demand for performance skyrocketed in the 1960's, all Chrysler had to do was dust off the Hemi design and apply it to the production big-block, which at the time displaced 426 cubic inches, safely under the racing limit.

It didn't take long for the effects to be felt. The Fords were practicing in the 165-167-mph range and the MoPars around the 175 mark.

Ford teams cried to NASCAR, but with Chevy and Pontiac gone, NASCAR wasn't about to throw out Ford's only competition, even though the Hemi wasn't production or even readily available over the parts counter, repeating Chevy's situation a year ago.

After resigning themselves to being slower again, the Ford partisans consoled themselves with the fact that last year's Chevrolets were also new and fast, but they didn't last.

Paul Goldsmith qualified a Plymouth fastest at 174.910, nearly ten miles an hour faster than the fastest Chevy the year before. A pair of front row qualifying races were held February 8 and Plymouths won them both, averaging over 170 mph.

Two hundred-mile races were contested February 21 to determine the rest of the starting field. A pair of Dodge drivers, Junior Johnson and Bobby Isaac claimed them. The best Ford drivers could do in either race was fourth, which Panch took in the first and A. J. Foyt nabbed in the second.

On Sunday, February 23, the Daytona 500 was held and those waiting for the MoPars to blow up had a long afternoon. The Hemis ran away from the Fords and were as strong at the end as the beginning. Led by Richard Petty, Plymouths took the first three positions. Jimmy Pardue was second and Goldsmith took third.

Sportiest of the 1964 full-sized Fords was the Galaxie 500/XL, here in two-door hardtop form. New styling for 1964 fastbacks brought new sheetmetal below the beltline with a familiar theme. Note the engine insignia, in this case for a 390, on the front fender.

Two laps behind, Panch brought the first Ford across the finish line in fourth. Fifth place Jim Pascal, in a Dodge, came in ahead of the first Mercury, driven by Wade.

Ford's Total Performance program didn't include getting its doors blown off, so reaction was immediate. Ford asked NASCAR to approve a limited-production engine of its own which put overhead-cam heads on the 427. Much to Ford's dislike, NASCAR said no, it wasn't production.

Work also began on a new engine package to get even more horses out of the regular 427. There was some development work on using Fairlanes as stock cars, but they wouldn't handle and major reworking was needed to shoehorn in the big engines.

MoPars didn't take everything after Daytona. They were heavier and did their best on big tracks where the power of the engines could be fully used.

Ford's counter blow, the 7000-rpm kit, actually worked very well. The earlier setup developed about 500-520 hp at 6200. The 7000 permitted higher revving and developed 550 at 6800.

A key to the higher revs was a lighter valve train with hollow stem valves. Also new were connecting rods with capscrews replacing the old nut-and-bolt system, better oiling with a cross-drilled crankshaft and a larger Holley carburetor rated at 850 cfm, compared to the old 780 unit.

The components of the 7000 kit were sold to the public through Ford's parts division.

Top Fords entered in the April 5, 1964, Atlanta 500 had the kits and they paid off immediately as Fred Lorenzen won in the Holman & Moody Ford. Fords won six of the next seven races. Lorenzen took the May 9 Rebel 300 at Darlington and suddenly the Hemi threat didn't look that bad.

Tragedy struck in the May 24 World 600 at Charlotte. Fireball Roberts was involved in an accident and his Holman & Moody Ford caught fire. Roberts was seriously burned and died six weeks later, just before the July Fourth Firecracker at Daytona. Foyt left Ford and won the race in a Ray Nichels Dodge. Lorenzen was injured and temporarily knocked out of action, and suddenly the Ford fortunes took a dive.

Holman & Moody tried several drivers, including an up-and-coming young driver Cale Yarborough.

Mercury started trimming its stock car program. Stroppe was told to cease his NASCAR operations and Darel Dieringer, his driver, was sent to Bud Moore.

When the 1964 season was over, Ford had won more NASCAR Grand Nationals than ever before, thirty. Dodge was second with fourteen, followed by Plymouth with twelve. Among the Plymouth flags were nine by Petty, who won the driving championship. Mercury, which announced its withdrawal from NASCAR racing at the end of the year, took five wins.

Despite the fact that Ford had its best season ever, there was still unhappiness over the Hemi. The months that followed started a period of politics and unrest in stock car racing that would blight the next couple of seasons.

Ford products made up the first four rows of the May 3, 1964, Yankee 300 on the road course at Indianapolis Raceway Park. On the pole is Fred Lorenzen (#28-1964 Ford) with Parnelli Jones (#15-1964 Mercury) alongside. Making up row two are A. J. Foyt (#2-1964 Ford) and an unidentified driver (#121-1964 Ford). The USAC stock race was open to NASCAR drivers. Lorenzen won the event.

Stroppe was told to also curtail his USAC stock operation during the 1964 season, but he convinced the factory to let him keep the equipment to finish it out —and finish he did! His driver, Parnelli Jones, won seven races, including three at Milwaukee and Pike's Peak for the second straight year, and ran away with the stock car driving title. Stroppe won the owner's title and Mercury the manufacturer's award.

Only two races were won by Ford, one of them the Yankee 300 at Indianapolis Raceway Park by visiting NASCAR star Lorenzen. Dodges accounted for four wins and Plymouth two.

Mercury left the stock car wars for a while to concentrate on the Comet but, at least as far as USAC was concerned, went out in a blaze of glory.

Another group that sanctioned stock car races in the Midwest, primarily on dirt was the International Motor Contest Association (IMCA). Dick Hutcherson of Keokuk, Iowa, drove a Ford to his second consecutive IMCA stock car crown in a Ford. Hutcherson had some factory help and after the 1964 season, went to NASCAR to seek his fortunes.

Also wrapping up his second straight title aboard a Ford was Jack Bowsher, who made a habit of dominating the Auto Racing Club of America (ARCA) circuit, which ran on paved tracks around its Toledo, Ohio, hub and a special race each February at Daytona.

In drag racing the torch in the hottest classes had been passed to smaller machinery. Fairlane Thunderbolts fought the battles in S/S, and A/FX contests were entered by Comet and Falcon compacts.

The Thunderbolts battled the MoPars on more even terms in the stick-shift classes, as indicated by Gas Ronda's win in the S/S class at Pomona and Butch Leal's similar performance in the NHRA Nationals at Indianapolis.

1965

In Detroit, the term "all new" has been overworked and undervalued over the years, being applied to cars with as little as a change of chrome trim, but when Ford said its 1965 full-sized cars were all new, it really meant it. The changes were the most drastic, especially underneath, since the 1949 models demonstrated Ford's new postwar styling and engineering.

Starting from the ground up, the frame and the concept were new. There were now coil springs front and back. The front units were redesigned for strength, utilizing the experience from stock car racing. Conventional coils were still between the upper and lower control arms. The design was so strong that it became the standard for NASCAR stock cars, regardless of make. Right into the 1980's, the cars may have Chevrolet, Buick or Ford sheetmetal, but up front the design is 1965 Ford.

The same thing later happened with the rear differentials. Ford units were and are considered the strongest, again thanks to racing development.

The rear coil springs were mounted just ahead of the rear axle with two control arms anchoring the axle and springs to the body. A third member was attached to the right-hand side of the differential. There was also a panhard rod from the right side of the axle to the left frame member.

All-new in styling and chassis was the 1965 Ford line. The new cars had revised suspension systems, completely restyled, smaller bodies and less weight. A good-looking example is this Galaxie 500/XL two-door hardtop, with its standard bucket-type seats and special trim.

Frames contained torque boxes for added strength. In addition, the bodies were strengthened similar to unitized bodies. The number of frame attachment points was reduced so the amount of road noise and stress transmitted to the body shell was reduced.

While this had little to do with performance, it did make for a quiet ride, something Ford would make a lot of noise about in the coming model year.

Bodies were also completely changed. The only link to the past was the two-door hardtop roofline, which was of similar proportions to the fastbacks of 1963½ and 1964. A close inspection revealed that even their sheetmetal was new. Four-door hardtops and sedans went back to a more formal roofline from the 1964 styles. Even the station wagons were redone. All models got curved glass in the side windows.

Vertical, dual headlights flanked the new three-dimensional straight bar grille. A primary straight ridge of sheetmetal ran along the sides of the cars and at the back were single, large hexagonal taillights in place of the traditional round ones. Base sedans did get round lights in the hexagonal housing however.

Chrome trim was kept at a minimum all the way around, with top line models' side trim being confined to rocker panels and wheel openings.

Probably the least changed were exterior dimensions. Wheelbase remained at 119 inches and front and rear tread moved up to sixty-two inches, from sixty-one in front and sixty in back. Overall length grew a mere 0.2 inch to 210, but width was down from eighty to 77.3 inches. Heights were within a fraction of an inch of the earlier models, depending on body style.

Reversing a trend, curb weights were down for 1965. For example, a 1964 Fairlane 500 two-door hardtop with base V-8 and three-speed manual transmission had a curb weight of 3,735 pounds. The same setup in 1965 tipped the scales at 3,539. Unfortunately, the efficiency drive would be a short one.

Reversing the trend to smaller wheels, Ford went to fifteen-inchers for all cars instead of just the 427-equipped models. Brakes remained eleven-inch drums, front and rear.

At the top end, the model lineup was changed, marking a major move of the big Fords toward luxury instead of performance.

Galaxie 500/XL's now shared top billing with a new series, the Galaxie 500 LTD. Available in two- and four-door hardtop forms, the LTD featured a plush in-

With most of the Chrysler products sitting on the sidelines, Fords dominated the 1965 NASCAR Grand National season. Starting on the pole for the Labor Day 1965 Southern 500 at Darlington was Junior Johnson (#26-1965 Ford). Alongside came Fred Lorenzen (#28-1965 Ford). Making up the second row were Marvin Panch (#21-1965 Ford) and Darel Dieringer in a 1964 Mercury. For many events the year-old Mercs were all the competition Ford had. Ned Jarrett won the race and helped Ford reach a NASCAR record of 48 victories in a single season.

terior with bench seats front and rear. Ford found out it was quieter than a Rolls-Royce and pounded that point home throughout the year.

Reduced to two models, the Galaxie 500/XL still did the bucket seat bit. Consoles and interior trim on the two-door hardtop and convertibles were redone. Buckets continued of the thin-shell variety. Exterior identification for the XL's was reduced to the nameplate on the front fender.

With all the attention the LTD's got, the 1965 models marked the beginning of a downward trend for the sporty XL versions.

XL production started with the partial model year run of 41,595 for 1962, peaked at 94,730 in 1963 and came close with 88,136 for 1964 models. The bottom dropped out for 1965 when only 37,990 were called for. This compared to 105,729 LTD's for 1965.

Overall, sales were up for the big Fords for the third model year in a row. A six-percent increase brought production up to 978,519. Returning to its former status as most popular model was the Galaxie 500 four-door sedan. Its tally of 181,183 outpaced the Galaxie 500 two-door hardtop by 23,899 cars.

Getting back to the nuts and bolts, the biggest change in the engine department for the big Fords was at the bottom of the lineup. A new 240-ci "Big Six" replaced the 223-inch version that had been around in basic design since the 1952 models. The new unit was advertised at 150 hp.

Elsewhere, the Challenger 289 V-8 gained five horses and, yes, the 390 still was good for 300 alleged horses.

Any model except wagons still could be had with the 427 and the usual associated heavy-duty goodies. Most literature did not mention the 410-horse single four-barrel version.

Actually, there were two versions of the 427 for 1965, the early and the late. Early 427's were much like the 1964 models, but after the first of the year, several changes were made, most coming from racing experience.

Available for 1965 was the spirited 427, one of three Thunderbird V-8 options this year.

Use of chrome trim was limited on the 1965 Ford Galaxie 500's all-new body. It has a badge on its front fender indicating a 352 under the hood. Sporty seats and spinner-style wheel covers were standard.

Biggest change was the block. A new, larger oil gallery for the main bearings was low, on the left-hand side. The bulge could be seen from the outside. It improved lubrication and became known as the "side oiler."

Also cylinder bores were scalloped so the maximum-size valves could be used. The nodular cast-iron crank was replaced with a forged-steel unit that was cross-drilled for better oil circulation, similar to the 7000 engines.

Also like the 7000 racing engines, there were new, lighter valves with hollow stems and thin, flexible heads.

Connecting rods were stronger and pop-up pistons were used. Heads had the combustion chambers machined for better flow and less danger of hot spots.

As usual, no increased output was advertised, but the 1965½ 427 was the strongest yet. The bottom end also served as the basis for Ford's single overhead cam V-8, which it would try to convince NASCAR to legalize and would sell through dealers to racers who used it in drag competition very successfully.

Despite the design change, three versions of the 427 would be available: the 410-horse NASCAR engine with single four-barrel, the 425-hp street version and the 425-hp drag engine.

In the middle of the 1964 model run, Ford brought out its own four-speed manual transmission, as demand was greater than Borg-Warner could supply. GM and Chrysler also came up with their own four-speed units.

Marvin Panch drove a 1965 Ford Galaxie for The Wood Brothers that year and accounted for four NASCAR wins.

Ned Jarrett became the first exclusive Ford driver to win a NASCAR Grand National driving championship in 1965. He did it in Bondy Long-owned Galaxies. Jarrett scored thirteen of Ford's forty-eight wins.

A. J. Foyt claimed his second straight Daytona Firecracker 400 trophy on July 4, 1965, driving a Holman & Moody Ford. Foyt won it in a Dodge the year before.

The Ford box was continued as an option for the 390 and 427 engines, carrying the same $188 price as the Borg-Warner T-10. Ford ratios were 2.32, 1.69, 1.29 and 1:1.

The inspection plate on the Ford four-speeds was on the top as opposed to the side on other makes. This gearbox became known as the "top loader."

Car Life tested a 425-horse street 427 in a Galaxie 500/XL two-door hardtop and with a 3.5:1 rear axle and four-speed. It claimed an amazing 0-60 time of 4.8 seconds, which was the fastest thing it tested since an A/FX Comet. The big Ford got up to 100 mph in 15.8 seconds and ticked off the standing-start quarter in 14.8 seconds at 97 mph.

While the big Fords could still hold their own on the street, not to mention in stock car racing, it was pretty well over for the full-sized Mercury. It too got a new chassis and body, but it also got some fat. Wheelbase went up to 123 inches, from 120; length and weight also grew. The racing program was dead and although you could still get a 427, it just couldn't be considered a performance car any more.

A mid-year change of engines for full-sized Chevrolets in 1965 proved to be a step in the right direction to bring that make back into the performance game after several years' absence.

Holman & Moody driver Fred Lorenzen copped four big wins for Ford in 1965, including the Daytona 500.

Features for 1965 included padded instrument panel, SelectAire conditioner, safety-convenience control panel and Studiosonic Sound System.

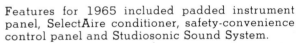

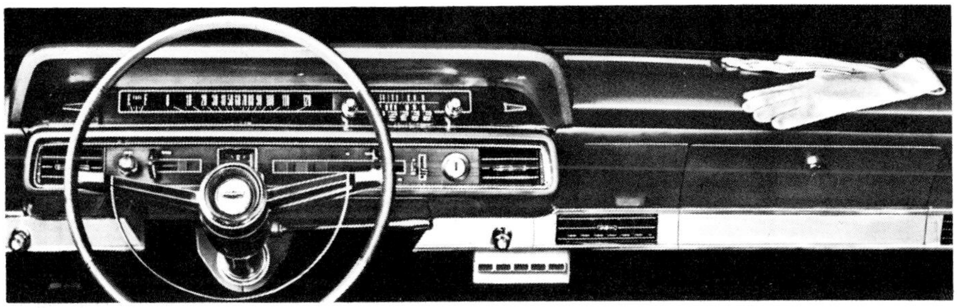

Though the roof line somewhat resembled the 1963½ hardtop, the 1965 Ford was new from road to roof. Even the side windows were curved instead of being flat. This Galaxie 500 two-door hardtop has a 352 aboard, as indicated by the badge on the front fender.

Replacing the aged 409 was a new 396-ci V-8 based on the Daytona Mark IV engine. At first it was offered only in the 396 size, but starting with the 1966 models it grew to 427 cubes and would give Ford a run for it in the big car performance sweepstakes.

Making its final appearance for 1965 was the lightweight Galaxie hardtop. Even the Fairlane Thunderbolts had given way to smaller Mustangs for Ford's front-line drag racing battles. Plans called for a few 427 Galaxies for B/FX and 289 Galaxies with Weber carbs for C/FX. The stock Galaxies were eligible for AA stock, but the competition was fierce there with older Super Stocks (notably MoPar wedges) with which to contend.

At the NHRA Nationals, Bud Shellenberger drove his lightweight AA/S Galaxie to the top stock eliminator crown, but it was a 1964 model, not a 1965.

Fords shattered the all-time NASCAR Grand National win record with forty-eight wins out of fifty-five events in 1965, but the explanation of the controversy that plagued the season is longer than the list of winners.

Briefly, NASCAR banned the Hemi and the Ford High Riser after Chrysler threatened to build a double-overhead-cam engine if NASCAR let the sohc Ford run. NASCAR then announced a 119-inch wheelbase limit (Dodge's and Ford's) for tracks longer than a mile and 116 (Plymouth's) for tracks a mile and under. A complicated set of specifications for three different classes was also announced.

This was too much for Chrysler and it halted its NASCAR participation and advised its drivers to do likewise. USAC was asked to go along with the Hemi ban, but didn't, so Chrysler said its teams would race there. Ford got mad and pulled its backing from its USAC team, Zecol-Lubaid.

Ford was strangely quiet about its High Riser being banned in NASCAR, because it had a new Medium Riser manifold ready to go that reportedly matched the High Riser in output.

The NASCAR season started with a nearly all-Ford show. The only real competition was the year-old Mercurys that Bud Moore was still fielding on his own. Gurney ran away at Riverside for his third Motor Trend 500 win, Lorenzen took the Daytona 500 and the crowds started to drop.

Negotiations with Ford, Chrysler, NASCAR and USAC produced a partial settlement by July. NASCAR agreed to reevaluate the Ford sohc for the 1966 season, let the Hemis run in smaller 1965 Plymouth Belvederes and Dodge Coronets on tracks a mile and less and in the newly introduced full-sized Plymouth Furys and Dodge Polaras on tracks longer than a mile.

Also imposed was a 9.36-pounds-per-cubic-inch rule, meaning a 427-powered car would have to weigh about 4,000 pounds.

Ford agreed to stay in NASCAR and return to USAC while Chrysler agreed to return to NASCAR for events in which it was legal. NASCAR also agreed to let Curtis Turner return to competition after his "lifetime" ban, which started in 1961.

Ford drivers had reeled off thirty-two straight wins before Richard Petty claimed a 200-miler at Nashville, Tennessee, on July 25 in his Plymouth. Two Ford drivers scored thirteen victories, Ned Jarrett and Junior Johnson. Newcomer Dick Hutcherson took nine for Ford and Marvin Panch and Fred Lorenzen each added four more. Jarrett won the driving championship, his second and the first for an exclusive Ford driver in Grand National history.

Ford's USAC campaign involved providing Holman & Moody Galaxies for Parnelli Jones and A. J. Foyt. Also a late-season deal was worked out with Zecol-Lubaid. Foyt took one win and independent Don White, driving a 1964 model, scored the other for Ford, otherwise it was a MoPar year with Plymouths winning a dozen events and Dodge two more. Plymouth factory driver Norm Nelson won the driving title. Jack Bowsher of Springfield, Ohio, repeated as ARCA champion, driving a 1965 Galaxie.

There were two versions of the big Ford lightweight in 1965, one with a 289 and the other with an sohc 427. Bill Hoefer stands on the former on his way to Junior Stock Eliminator honors in the 1965 NHRA Springnationals at the Bristol (Tennessee) Dragway.

It should be noted here that among the non-factory teams and short-track drivers, there was a resistance to the 1965 big Fords. Many preferred to stay with the proven leaf spring rear suspension on the older models.

The 1965 big Fords provided an interesting paradox. They were probably the fastest production full-sized car available when equipped with the 427 High Performance package. However for the average buyer who wanted automatic transmission, power steering, power brakes, air conditioning and the other gingerbread, all Ford had to offer was the sedate 300-horse 390.

1966

A trend among full-sized cars at the time was the low-stressed, big-cube engine that could run all the accessories, provide good acceleration and highway performance and keep quiet about the whole thing.

An example was Chrysler's new 440-ci wedge for its full-sized cars that was said to produce 365 hp. The powerplant wouldn't be called upon for performance duties until 1967.

This sports-type steering wheel for 1966 had a walnut-like rim.

Ford's instrument panel for 1966 was designed to be good looking and practical. "Swept-away" panel design created extra knee and legroom; steering effort was reduced up to 85% with the power unit.

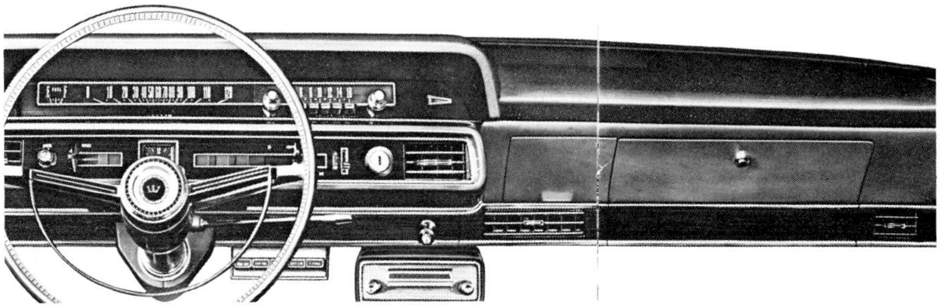

New to the 1966 Ford line was the 7 Litre series, consisting of two-door hardtop (shown here) and convertible. A 345-hp 428 was standard, as was striping, special trim including side striping and bucket-seat interior.

Ford answered the call with a new engine and a new series of models. By using the bore from its 406 and the stroke from a 410-ci engine Mercury offered (4.13x3.98), it came up with the 428. Actually the displacement worked out closer to 427 cubes, but the 428 number was used so it couldn't be confused with Ford's 427.

It was rated at 345 hp at 4600 rpm and had features like hydraulic lifters, a 10.5:1 compression ratio and single four-barrel carb. Like the 390, it could be equipped with all the power stuff.

Missing things like mechanical lifters, pop-up pistons, big valves and headers weren't noticed. An Interceptor version for the police was also available, rated at 360 at 5400. It was the first break from the 330-horse 390 for the police package since the first one in 1961. The police version had an aluminum intake manifold and hydraulic lifters.

The new 345-hp engine came standard in the Galaxie 500 7 Litre series and optional on other models, including station wagons.

Standard equipment on 7 Litre models (a two-door hardtop and convertible) also included Cruise-O-Matic, the new power front disc brakes and an XL-type interior with bucket seats and console. Optional powerplants were the 427's, now called Cobra V-8's.

Other than the new series, the rest of the 1966 big Ford model lineup remained the same. Styling was a facelift with the grille split into two horizontal sections. Side sheetmetal was altered slightly with a subdued kickup for the rear quarter panels, which added 1.7 inches to the overall width, to bring it to seventy-nine inches. Taillights were now squarish and the less expensive Custom and Custom 500 models no longer got round versions.

In model nomenclature, the LTD was no longer considered a Galaxie 500, but the XL was.

For cars with 428's, two new driveline features were available, Ford's new C-6 automatic transmission and a larger 9⅜-inch ring gear for the differential. Both were strong units and would see use in other Ford performance cars during the decade.

Cobra 427 V-8's were still available with the High Performance package and still only came with four-speed manual gearboxes. They weren't available on wagons and continued to be advertised at 410 and 425 hp.

This 7 Litre emblem of grille and fenders hinted at what was under the hood of a 1966 Galaxie 500—the mighty, new Thunderbird 428 V-8. It was 345-hp strong, with four-barrel carburetor, hydraulic lifters and special dual, low-restriction exhaust.

Most popular model in the 1966 full-sized Ford line was the Galaxie 500 two-door hardtop. Like other styles, it featured facelifted styling. Orders reached 198,532 to put it ahead of the runner-up Galaxie 500 four-door sedan with 171,886.

New for 1966 was a redesigned intermediate Fairlane, which could hold big-blocks, including the 427, under the hood. That was proven with a mid-year offering, making the need for such a high-performance engine in the big Fords questionable.

Car Life tested a 1966 7 Litre hardtop with automatic and found it could sprint to sixty in eight seconds flat, which wasn't bad for a car with a test weight of nearly 4,500 pounds.

New car sales during the 1966 model year reached an all-time high up to that point. Production hit 8,606,166, a figure that wouldn't be passed until 1973. Big Fords continued to gain, going over the million mark for the first time since 1959, with 1,034,928 being built in the U.S. The figure represented a 5.76-percent gain and the high-water mark for the decade.

Production of the XL's continued to tumble, dropping to 32,075. The 7 Litre wasn't all that successful, picking up 11,073 orders. Once again, the Galaxie 500 two-door hardtop was the most popular, downing its series mate four-door sedan, 198,532 to 171,886.

The 1966 full-sized Ford would be the last with any significant factory support in stock car racing, and that only lasted half a season. Some Galaxies were drag raced, like Mike Schmitt's sohc 1966 Galaxie, which did well; but in general, Mustangs now held the spotlight.

The Black Pearl was a 1966 Ford LTD-based show car that made the rounds showing how clean a little customizing could make a two-door hardtop look. Changes were made front, side and rear.

An extremely successful Ford Galaxie driver in drag racing was Mike Schmitt, who is shown at the 1966 NHRA Springnationals at Bristol, Tennessee, on the way to Street Eliminator honors in his 1966 sohc-powered Galaxie 500/XL hardtop. He ran in B/FX.

Making the rounds at auto shows in 1966 was the Magic Cruiser, which was based on the 1966 Ford Galaxie 500. It could be converted from a fastback to a station wagon, all by power controls.

Jack Bowsher switched to USAC stock competition and finished seventh in points, the highest for a Ford driver on the circuit. For the first time in USAC history, a Ford didn't win a stock car event during the season. Norm Nelson scored seven of Plymouth's nine victories and took a second straight title. Don White switched to Dodge and scored all eight of its USAC wins to place second.

Then there was NASCAR.

Ford made another attempt to get the sohc 427 legalized, but NASCAR said it wasn't a production engine yet. Ford later said it was a standard, over-the-counter item with a price tag of $1,963 (compared to $900 for the Hemi). There were rumors of the sohc going into a limited number of Galaxies, but the production never took place.

To make Ford's case weaker, Chrysler was now offering a street version of the Hemi in its intermediate Plymouth Satellite, Dodge Charger and Coronet lines.

As was tradition, Gurney romped at Riverside, leading 102 of the 185 laps in his Wood Brothers 1966 Galaxie.

Factory-backed drivers in Ford's abbreviated 1966 NASCAR Grand National season in their 1966 Galaxies are Cale Yarborough (#27) driving for Banjo Matthews, Curtis Turner (#41) driving for The Wood Brothers and Fred Lorenzen (#28), in a Holman & Moody car. Ford teams sat out a good part of the season over a rules dispute and when they returned, Fairlanes took over as the standard car.

Though in the shadow of the luxury-oriented LTD, Galaxie 500/XL continued for 1966. This convertible has the standard bucket-seat interior and plaque on the front fender indicating a 390 V-8 is aboard.

At Daytona it was a different story as Richard Petty outclassed everyone after a year's absence and took his second win in the 500.

Some Fords used specially contoured bodies, which looked somewhat like bananas, with drooped noses and kicked-up tails. The MoPars were still faster.

Finally, after an appeal to the Auto Competition Committee of the U.S. (ACCUS), the sohc was declared legal for installation in Galaxies. Only problem was, a penalty of one pound per cubic inch was assessed against the sohc cars. That meant sohc Galaxies would have to weigh 4,424 pounds compared to 4,000 for the Hemi cars.

This was unacceptable, and on April 15 Henry Ford II announced his company was withdrawing from NASCAR competition. This time the Fords stayed home and the MoPars romped, the opposite of the previous year.

Parallel to the previous year was a July reconsideration of a return. Ford announced it would be back on a limited scale.

While the big guns of Ford were away, a major change took place. Bud Moore, now independent, built a 1966 Mercury Comet, which was a twin to the Fairlane, and got the thing working, proving the intermediate Fords could now handle the tough stock car racing chores. Other factory teams built Fairlanes and when the Fords returned, one by one, they switched to Fairlanes.

Moore's Comet driver Darel Dieringer won the September 5 Southern 500 at Darlington and victories followed by Fred Lorenzen in a Holman & Moody Fairlane, including the season-ending American 500 at the North Carolina Motor Speedway in Rockingham.

While individuals continued to build Galaxies for racing for many years to come, they no longer were in the factory's plans for racing involvement.

Just as Ford drivers in 1956 and 1957 prided themselves in knocking off the bigger cars, now Galaxie drivers were suffering the same fate at the hands of smaller-machinery pilots.

Despite this turn of events, Ford just didn't fold up its sports accents for its full-sized models. XL's continued in the lineup through the 1970 offerings and the 427 was carried on the option list into the 1968 model year.

1967

Though it didn't make much difference, as the competition was doing it as well, the big Fords got bigger, gaining pounds and inches on a yearly basis. To indicate its direction, the press release for the 1967 big Fords said, "The 1967 Ford will seek a larger share of the medium price market by offering the biggest and most luxurious cars in its history." It went on to proudly proclaim three inches of additional length. Actually Ford must have forgotten the 1960 models, which were both longer and wider.

XL two-door hardtops were of a fastback design that was shared with the Galaxie 500. However it wouldn't make much difference on racetracks as the hottest Fords for 1967 in stock car racing were Fairlanes. From the side, identification for the XL was nil. The emblem on the front fender indicated the engine size, which in this case was a 428.

Sheetmetal for the sedans, hardtops and convertibles was new and there were now two two-door hardtop designs, a fastback version with smoother lines than the 1966 models, and a formal-roofed affair. The fastback came in Galaxie 500 and XL versions, while the other was available as an LTD.

Standard on the XL's were redesigned buckets, a console, the 289 Challenger V-8 and SelectShift Cruise-O-Matic. An option was the 7-Litre Sports Package, which included the 428, power front disc brakes, a heavier suspension system, wide oval tires and a simulated wood steering wheel. Cost of the package was $515.86.

Other than the LTD getting a four-door sedan, the model lineup for 1967 remained the same. The engine lineup was a carryover with the 427 being a limited production option and coming with and without the things it did in past models.

Continuing its sales trend, demand dropped again for the XL models, reaching a new low, 23,335, for the model year production. The fastback hardtop wasn't to blame, for in Galaxie 500 form, it was the most popular of the full-sized Fords with 197,388 being made domestically, well ahead of the 130,063 tally for the Galaxie 500 four-door sedan.

Overall, production of the 1967 big Fords was off a little over fifteen percent, compared to an industry slowdown of nearly eleven percent.

Another set of new sheetmetal marked the 1967 full-sized Ford. The new models looked longer and were, gaining three inches in length. Looking spiffy was this XL convertible with optional styled steel wheel covers.

Ford went back to two different two-door hardtop designs for the 1967 model run. The popular LTD series got a formal roof while Galaxie 500 and XL models got a fastback style.

1968

Significant in the performance history of the big Fords was the 1968 model year, not for what was gained, but for what was lost. Actually things got worse as it progressed. The main links to the Total Performance days, the 427 and XL, started their 1968 model life in diluted form.

Early in the model year, the 427 was carried on the option list, but it was not the 427 of old. The size was the same, but with the new federal emission standards, it had to be toned down.

Called the Cobra V-8, it was rated at 390 hp at 5600 and torque worked out to 460 pounds-feet at 3200, both ratings milder than previous street editions. A single four-barrel carb was the only manifolding available, compression was cut to 10.9:1 and for the first time, hydraulic lifters were installed, with the oiling system revised to feed them.

Another first was the attachment of the 427 to an automatic in production big Fords. In fact the only way you could get the 427 for 1968 was bolted to a C-6 Cruise-O-Matic.

Fixed horizontal headlights were a feature of the Galaxie 500, here in convertible style, while XL and LTD models received hidden headlights. The face-lifted cars were the last with a 119-inch wheelbase. This example is equipped with a 390, if they bolted on the right engine emblem, which is on the front fender.

Both fastback and notchback versions of the two-door hardtop were available in the Galaxie 500 series for 1968. Before, only the LTD got the formal style. Notchbacks, like the one shown, were more popular in 1968 than the fastback, despite having a slightly higher price.

The 427 was again available in any model except station wagons. With it came a stronger 80 amp-hour battery, power front disc brakes, G70x15 wide-oval tires, heavy-duty suspension and heavy-duty 3.25:1 differential.

The 427 was also offered in Mustangs and Fairlanes and was replaced mid-year by a performance version of the 428, the Cobra Jet. This move ended the career of the 427 as a production line option. It continued to be offered through Ford's parts operation. The big Fords did not list the Cobra Jet as an option.

At its peak, Ford's 427 was one of the most powerful production engines ever made for a domestic passenger car. The only one clearly stronger was the Chrysler second-generation Hemi. Chevrolet's big-block, which reached its zenith in the 1970 models at a size of 454 ci could rival in raw power the output of the hottest 427's, which were made a few years before.

Undergoing changes in its identity was the XL series. There were two versions, fastback two-door hardtop and convertible.

At first bucket seats and a console were standard and a front bench seat was optional. This was changed around shortly into the model run with bench seats now standard and the buckets an option.

Many changes marked the 1968 Fords. New was a more pronounced fastback for the XL and Fairlane 500 series. It was the first year for the XL that a V-8 wasn't standard and mid-year XL lost its standard bucket-type seats and console. This XL fastback has the optional GT Equipment Group with stripes, simulated mag wheel covers, GT badges on the front fenders and heavy-duty undersides.

Starting with the 1967 models, full-sized Fords became rarer in stock car racing, but some were still built through the late 1960's, usually on older racing chassis. Posed with the 1968 Ford fastback he raced in USAC is Indianapolis car driver Art Pollard.

One favorite trick to create Ford show cars in the late 1960's was to chop the roof. Such was the case with the Fiera, based on a 1968 XL fastback. Ford Design Center in Dearborn did the job. The car appeared at auto shows during the year.

The resulting move was a reduction in price from the 1967 models. Helping the price cut was changing standard equipment from a small V-8 and automatic to the six and three-speed manual transmission.

Prices for the 1967 XL's were $3,243.24 for the hardtop and $3,492.84 for the convertible. In February of 1968, these had dropped to $3,040.65 and $3,270.18, respectively.

The change did not go unnoticed by the buyers, as the XL series attracted 56,114 orders, an increase of 140 percent and the highest since the 1964 models.

It's not that the 1968 was without redeeming high-performance value. Provided your XL had at least the 265-hp 390, you could plunk down $204.64 and get the GT Equipment Group. That meant the heavy-duty suspension, low-restriction exhaust, GT emblem and stripes along the side, wide-oval tires, power disc brakes and genuine simulated mag-type wheel covers.

Overall, the full-sized 1968 Fords were restyled from the beltline down, except for the two-door hardtops, which got new fastback and notchback roofs. Length went up 0.3 inch and width actually came down an inch to seventy-eight. The LTD got the notchback hardtop, the XL the fastback and the Galaxie 500 both.

Other than the 427 deletion, engine lineup remained much the same as in 1967, with the advertised outputs being juggled a few horses on the smaller engines. The four-speed manual gearbox was still an option on the 315-hp 390 and 340-hp 428. Police cars still got an Interceptor version of the 428, rated at 360 hp and available only with the C-6 Cruise-O-Matic.

New car production for the 1968 models was up 9.6 percent. Big Fords did not share in the prosperity, as assembly lines cranked out 867,292 for a little over a percentage point drop; but a strike early in the model year no doubt had a major effect. Galaxie 500 four-door sedans were the most called for with 117,877 being built. The notchback Galaxie 500 hardtop was a weak second at 84,332.

1969

Big Fords got bigger for 1969 with wheelbase rising to 121 inches, the first increase since 1960 models. Length varied with models, but on the XL and LTD versions, it reached a record 216.9 inches. Tread also grew, hitting sixty-three inches in front and sixty-four in back.

For 1969, XL hardtops got a new roof line and a new name, SportsRoof. Though it looked like a fastback from the side, the effect was created with a notchback rear window and sail panels. As in the late 1968's, bucket seats and a console were optional on XL's.

Sportiest of the 1969 full-sized Fords was an XL convertible with the GT Performance Group. The package included mag-style wheel covers, GT letters on the side trim on the front fenders and C-stripe on the front. Bigger tires and competition suspension were also included.

Back again was the watered-down XL series with yet another newly designed two-door hardtop and convertible. Optional again was the GT Performance Group. But the bucket seats and console were not even exclusive XL options any more, as shortly after the start of the model run they could also be had on Galaxie 500 two-door hardtops and convertibles.

Body styling was all-new and interiors were redone. Two designs of two-door hardtops continued with the formal and the SportsRoof being offered. The latter looked like a fastback from the side, but long, tapered sail panels trailed off from either side of the notchback rear window.

Engine lineups changed for 1969 with a new "385" series big-block 429 topping the list. The Thunder Jet V-8 was available in 320-hp two-barrel form and 360-hp four-barrel. The latter engine was the only one on which the four-speed manual transmission was optional. Police cars still got the 428 Interceptor rated at 360 hp.

Mid-year big Fords got a badly needed mid-range engine, the 351 small-block. Based on Ford's 221-302 family, it filled the gap between the 302 and 390. It came in a 250-hp two-barrel-carbed version.

While new car production for 1969 was up only slightly from a year earlier, the 1969 Fords were popular with big car customers. The all-new styling resulted in production being increased seventeen percent, resulting in a cracking of the million mark, 1,014,750. It would be the last time the big Fords hit seven figures. XL orders climbed 10.4 percent to 61,959.

An option for 1969 XL and Galaxie 500 two-doors was the combination of bucket-type front seats and a console, here in XL trim. New that year was a cockpit-styled instrument panel with all the controls in front of the driver.

Formal roof two-door hardtops were available in the LTD (shown) and Galaxie 500 lines of the full-sized Ford for 1969. Wheelbase grew to 121 inches.

We're dealing here in remnants of the bygone days when big Fords could run with the best performance cars of the hour. Ford marketing and even the majority of the big Ford customers did not care about the heritage or even the performance of the late 1960's cars. All were far more interested in loaded LTD's rather than four-speed, heavy-duty-suspended XL's.

Car Life tested an LTD four-door hardtop with the big 429 and standard suspension. Its curb weight was 4,565 pounds and its test weight approached 4,900. It also tested other full-sized cars and commented, "Its [Ford's] handling put it at the bottom of the group, despite its radial tires. It took nerve to take this car into the constant radius turn much faster than 40 mph."

1970

Our final look at a big Ford model will be the 1970 edition, and that is only because the XL designation was still used.

For once, there were only minor trim changes for the Ford giants. XL's got the 250-hp 351 as standard equipment, but buckets and a console were still an option, shared with Galaxie 500 two-doors. The GT option disappeared, as did the choice of a four-speed transmission.

About the only option worth noting was Dualtone paint, which involved a flat-black hood, cowl and fender tops. Flat-black paint then extended in a narrow stripe under the windows to the rear roof pillar.

Ford's performance *Buyer's Digest* for 1970 showed an XL convertible, saying "Here's XL, big daddy of Ford's fun fleet," then went on to call attention to its trailer-towing ability.

XL production declined nearly forty-six percent to 33,599, the line was down nearly sixteen percent at 852,992. The industry production was off about ten percent.

Emphasis on performance was cut back after the 1970 model year by Ford and its competitors. Insurance companies had made the really powerful cars expensive to cover, safety and emission standards increased weight and reduced performance, and the threat of many more regulations on the horizon sapped the performance budgets of the manufacturers.

Not only that, but buyers were tiring of screaming tires and thirsty engines, and leaning more toward luxury. In its 1970 model run, Ford unloaded 265,020 LTD's, nearly eight times the number of XL's.

There were no more XL's after the 1970 models. Ford, like its competitors, concentrated on luxury and standard transportation. Ford convertibles were made through the 1972 model run and ended. Two-door hardtops vanished after the 1974 models.

1979-1987

The big Fords were downsized for the 1979 model year to a 114-inch wheelbase and the basic car remained in production relatively unchanged through the 1987 model year. Performance applications for full-sized cars were very much a thing of the past. The closest the big Fords came was in the police packages, which were not available to the public in new-car form. Available was the only passenger-car use of the 351 in recent years. Reports in late 1986 indicated that the 351 could return to the Mustang, however.

Mild restyling marked the 1970 full-sized Ford. The formal-roofed LTD with the Brougham option contrasted with the lines of the SportsRoof.

Ford At Le Mans

Yet another facet of the "Total Performance" sprawl of Ford in racing was the GT cars. A crew of talented people was assembled to work on the project with the goal of victory in international endurance sports car racing.

The first result was the GT-40, which reportedly got the number because it was forty inches high. Construction was completed in England and power was intended to come from Ford's Indy ohv 255-cid engine. The 289 was substituted and by late 1964, ten examples had been built.

The cars were not winners, so the next phase was to turn the racing operations over to Carroll Shelby, who went to 427 power for a Mark II version. Results came quickly and a Shelby car co-driven by Ken Miles won the 1966 Daytona Continental. The big victory came in June of that year when Mark IIA's finished first, second and third in the twenty-four hours of Le Mans, becoming the first American car ever to win. After a controversy, it was ruled that the team of Bruce McLaren and Chris Amon were the winners.

To top it off, Ford won the manufacturer's championship in the prototype category in 1966, beating Ferrari.

In 1967 the redesigned GT Mark IV took over and was an immediate success with McLaren and Mario Andretti co-driving to victory in the Sebring twelve-hour contest and A. J. Foyt and Dan Gurney teaming to repeat at Le Mans. There were two GT teams, one from Shelby and one from Holman & Moody. Gurney and Foyt drove for Shelby.

After Le Mans the FIA (Fédération Internationale de l'Automobile), which set the rules, announced that the big engines would be banned from the prototype cars, but the decision didn't upset the Ford folks all that much. Henry Ford II was there to witness the all-American victory, and that to him was satisfaction enough. He'd beaten Ferrari and the best that Europe had to offer—at their own game and on their own turf.

A street version of the GT was offered for a while before production ceased in 1968.

Ford's entrant in international endurance sports car racing was the Ford GT. This 1967 Mark IV won both the 12 hours of Sebring and 24 hours of Le Mans in 1967. It was powered by Ford's 427.

CHAPTER FIVE

Intermediates In Action Fairlane, Torino And Montego

As the 1950's progressed, so did the dimensions and weight of the so-called standard-sized domestic car. With the dawn of the 1960's came a group of compact cars from the Big Three.

Between the compacts and the big cars was a gap that approximated the size of the standard low-priced cars in the early 1950's. They had wheelbases in the 115-inch range and lengths a little under the 200-inch mark.

The closest thing in the market in 1961 was the 117-inch-wheelbase Rambler Ambassador and its shorter, more popular cousin, the 108-inch-wheelbase Rambler Classic.

After the compact car onslaught, the next battleground for the Big Three was this "intermediate" segment of the market.

First in was Ford with its 1962 model Ford Fairlane and Mercury Meteor.

Ford's approach was an extension of its compact car engineering, which provided for a lightweight, efficient package. Its competitors that followed used a different approach, extending their big car engineering to the class.

While the Ford intermediates may have been fifteen years ahead of their time in the efficient engineering concept, they ran into an unexpected problem when big-engine-equipped intermediate cars became the hottest segment of the booming performance car era, which was just down the road.

In time Ford adapted and turned out some of the hottest intermediates of all, but the transition was not easy.

Ford couldn't come out and say it was aiming at Rambler with its 1962 Fairlane, so its published target was the 1949 Ford, their first all-new postwar car. The early Ford had a 114-inch wheelbase, overall length of 196.8 inches and width of 71.7. It carried six people in relative comfort. Four-door sedan curb weight came to 3,140 pounds. The Fairlane package had a 115.5-inch wheelbase, length of 197.6 inches and width of 71.3. A similar sedan tipped the scales at 2,904.

In comparison to its family members, the 1962 Fairlane was about 500 pounds heavier than the compact Falcon, but half a ton lighter than the full-sized Galaxie. Length worked out to 16.5 inches longer than the Falcon and 11.7 shorter than Galaxie. So, it may have been closer to the Galaxie in size, but it was much closer to the Falcon in weight.

Fairlanes followed the Falcon's unit construction and suspension configuration. Coil front springs above the top A-frame in front and leafs in the rear were similar on both cars.

The Fairlane's unit body was reinforced with a new "torque box" design which placed small boxes of steel in back of the front wheels and ahead of the rear to absorb shock and torsion not taken care of by the springs. The stress was then passed on to a member under the rocker panels. The front torque box was right under the toe board, which unfortunately transmitted some of the noise into the passenger compartment. That would be changed a few years later.

The reinforced understructure and relatively light weight made for a solid, smooth-riding car, capable of economy rivaling the compacts.

Styling was reminiscent of several recent models. Dual, horizontal headlights were in the grille, which looked like the 1962 Galaxie. Rear fins flared somewhat like the 1957 Ford and side trim on the top-line Fairlane 500 was like the 1959 Fairlane and Galaxie full-sized cars. Big round taillights continued a Ford tradition and rear styling was not unlike the 1961 full-sized cars. Roof styling was traditional Ford formal.

Even the name, Fairlane, came from the big cars of the past. Originally the name of Henry Ford's estate, Fair Lane, it was first used in production Fords as a name for the top-line 1955 models and changed to one word, Fairlane.

Starting with the 1957 models, top-liners were called Fairlane 500. A mid-1959 model dropped the Fairlane 500 designation to mid-range, being supplanted by Galaxie. The Fairlane name was last used to designate the lowest-priced full-sized 1961 Ford. Fairlane 500 was the middle series. As the Fairlane brochure said, "Some cars have new names . . . this name has a new car."

Initially, four models were offered, the base Fairlane club and town sedans (two- and four-door) and the fancier Fairlane 500 club and town sedans. The model line would grow many times in future years.

Standard engine was the 170-ci, 101-hp in-line six that was an option for Falcons. However, it was the optional engine that attracted more attention than any other feature on the Fairlane, a 221-cube V-8 of an entirely new design.

At the time, there was much controversy over the engine material of the future. General Motors, American Motors and Chrysler all brought out production engines with aluminum blocks. The lighter weight was considered an advantage, but it was balanced by higher production costs and an unexpectedly high scrap rate due to casting problems.

Ford did not follow the aluminum trend, but rather sought to improve the casting methods with traditional iron. Cast-iron blocks of the day were thick and heavy to allow for core shifts in casting. By precision casting, Ford found it could use thinner walls and save weight, and still have a strong block.

The first production application of the technique was the Falcon six. It was successful and the new small-block V-8 was the next major test.

Not only was the new Fairlane V-8 block light (it weighed 120 pounds compared to 144 for the Chevrolet 283), but it was compact. It measured twenty inches wide across the exhaust manifolds, twenty-nine inches from the tip of the front water pump to the back of the bell housing and twenty-eight inches from the top of the air cleaner to the bottom of the oil sump.

At the time, the Chevy small-block was the standard for small V-8's and it was twenty-seven inches wide and weighed about 535 pounds complete, ninety more than the new small Ford.

One area where weight was saved was in block design. Since the 1954 models, Ford overhead-valve V-8's came with Y-blocks, which extended the skirt past the centerline of the crankshaft. The thinking was to make a stronger bottom end. That philosophy changed on the new V-8 and the block only went to the crank centerline, as on most other makes.

Precision casting was also used on the heads to keep weight down. Also adding to lower weight was the use of studded rocker arms, similar to those on the Chevrolet V-8's. Previous Ford ohv designs used rocker arm shafts, which added

Fairlane started life as an intermediate-sized car with an efficient design and styling reminiscent of earlier Fords. This 1962 Fairlane 500 four-door sedan held down the top-of-the-line spot that year.

weight. While Chevy used stamped rocker arms, Ford utilized cast ones. Hollow pushrods carried oil to the top end and had a ball bearing on the end for durability. Hydraulic lifters were used and valves were a small 1.59-inches intake and 1.39-inches exhaust.

Since there were only 221 inches to feed, thanks to a 3.5-inch bore and 2.87-inch stroke, a small 210-cfm Ford carburetor was used with a cast-iron intake manifold.

The original rating was 143 hp at 4500 rpm, but that was changed after introduction to 145 at 4400. Torque was 217 pounds-feet at 2200 and later changed to 216 at the same rpm. Compression ratio listed at 8.7:1.

What impressed observers was the room to grow. The bearing area seemed far stronger than the original cubic inches warranted. The size of 221, by the way, wasn't just picked out of the air. It was the size of the original Ford V-8 introduced in the 1932 models.

Finishing out the Fairlane powertrain was the choice of three transmissions: three-speed manual, three-speed manual plus overdrive and two-speed Ford-omatic. Strangely, the overdrive was only available with the V-8. Axle ratios ranged from 3:1 to 3.5:1. Showing the compact car heritage, wheel sizes were the same diameter as Falcon with thirteen-inchers standard. An option was the more sensible fourteen-inch units.

Performance for the Fairlane V-8 wasn't outstanding. *Motor Trend* reported a Fairlane 500 two-door sedan with V-8 and automatic capable of a 0-60 sprint of

Fairlane's first mid-year model was the 1962 500 Sports Coupe, which had bucket-type front seats and a mini-console. The main exterior identification was the nameplate on the right side of the trunk lid. The emblem on the front fender, behind the wheel, indicates this model had the mid-year Challenger 260 V-8.

Introduced along with the Fairlane for 1962 was Mercury's intermediate, the Meteor, which shared the same basic body and engines, but had larger exterior dimensions. The most interesting view of the 1962 Meteor was the rear, which looked like its bigger brother, the Mercury Monterey. This Custom two-door sedan started off the model year as the best-trimmed, but it lost that title mid-year when the S-33 came out with bucket seats.

13.3 seconds and quarter-mile performance of 20.8 seconds with a trap speed of 70 mph.

While the Fairlane and Meteor had no direct competition in their class, they did get some unintentional pressure from Chrysler Corporation. It downsized its full-sized Plymouth and Dodge Dart lines for 1962. Under all that strange-looking sheetmetal was an intermediate-sized body and chassis. Both had 116-inch wheelbases and overall lengths just on the other side of the 200-inch mark.

At first, 1962 engines were cut back with the largest being 361 cubes, but pressure from the racers resulted in the release of big Chrysler engines up to 413 cubic inches. With less weight, a smaller package and a big engine, the MoPars began to flex their muscles in racing, especially in the drags where they were blowing the big 406 Galaxies into the weeds.

Ford was still intent on a sensible Fairlane package, but it didn't take long for the first performance option to come along.

Not only was performance a hot item early in 1962, but so were bucket-type seats, which nearly every make had. Both needs were taken care of with the 1962½ models that were introduced as part of Ford's "Lively Ones" campaign.

Joining the Fairlane lineup was Fairlane 500 Sports Coupe with front bucket-type seats and mini-console which was similar to the Falcon Futura, which came out the previous year. Along with it the Challenger 260 V-8 bowed, marking the first of several displacement increases for the new small-block.

While mid-year introductions of more powerful engines in some car lines coincided with race-sanctioning bodies' deadlines, the 260 introduction was timed to add spice to the Fairlane, as the car was not a contender in any major form of automotive competition at the time.

To arrive at 260 cubes, the bore was upped 0.3 inch to 3.8, while the short stroke remained the same at 2.87. Advertised output was a horsepower rating of 164 at 4400 and torque of 258 at 2200. Carburetion was still two-barrel and compression a regular gas demanding 8.7:1.

Besides the Fairlane and its cousin, the Meteor, 260's were also going into experimental Falcons, Mercury Comets and the Cobra project of former sports car racer Carroll Shelby, who inserted it into the vintage British AC sports car. Also, the engine would be adapted to power new rear-engine cars for the Indianapolis 500 being developed by Lotus in England.

Ford's new small-block may have been fairly calm in production form, but its performance potential was being explored on many fronts, in and out of the company.

Checking out the 260-powered Sports Coupe, *Car Life* noted, "The 'lively' engine in the lively-appearing Sports Coupe makes an appealing combination." It peeled off a 12.1-second 0-60 run, which was 3.4 seconds faster than an earlier test of a 221-powered four-door sedan.

The Sports Coupe may have improved the image of the Fairlane, but didn't do much for sales, as the 19,628 produced accounted for only 6.6 percent of the 297,116 made the first model year. Most popular was the Fairlane 500 four-door sedan at 129,258.

A new two-door hardtop design joined the Fairlane lineup for 1963. It came in Fairlane 500 and Fairlane 500 Sports Coupe, illustrated here. Among the exterior features for the Sports Coupe were roof emblem, portholes in the side trim and simulated knock-off wheel covers. The badge at the front of the front fender indicates the 260 V-8 powered this car.

While nearly 300,000 orders made the Fairlane a success, the same couldn't be said for the Meteor. It used the same basic body, had an inch more of wheelbase and 6.2 inches more length, but the same powerplants. Mid-year it also got the 260 and a bucketed coupe, the S-33, but it didn't click with the buyers.

Only 69,052 Meteors were ordered for 1962 and 1963 was even worse, despite the addition of a two-door hardtop and wagons, as only 50,775 went out the door. Like its celestial namesake, it didn't last long and was canceled after the 1963 model run.

1963

A minor facelift and new body styles distinguished the early 1963 Fairlanes. Performance characteristics would be enhanced as the model year progressed.

Expanding the lineup were two-door hardtops and four-door station wagons. The formal-roofed hardtop came in Fairlane 500 and Fairlane 500 Sports Coupe versions, the latter replacing the two-door sedan 1962 edition. Wagons came in two-seat Ranch Wagon and Squire versions with third seats optional.

Sport Coupes for 1963 featured all-vinyl buckets with a choice of five colors. Consoles now extended to under the dashboard. Rear seats were bucket-styled benches. On the outside was special trim with three Buick-like portholes on the front fender and a red, white, blue and chrome insignia on the rear roof pillar. Also standard were deluxe full-wheel covers with simulated knock-off spinners. A mid-year option was spoke wheel covers with knock-off hubs, simulated of course.

It should be noted that Chrysler was now selling 426-cube wedge engines for its Fairlane-sized Plymouths and slightly larger Dodges.

While all the simulating was going on, Ford engineers were working on some genuine improvements for Fairlane performance, within the confines of the package.

First out the door was an all-synchro three-speed manual for V-8 cars. Galaxies also got the new Ford-built box and while it didn't do much for acceleration, it did make the Fairlane more versatile for road work.

Options for 1963 included fully chromed simulated wire-wheel covers and top-mounting-style electric clock.

The 1963 Fairlane dash—attractive and convenient.

Next at bat was the Borg-Warner T-10 four-speed manual box, which was common on the option lists of many makes at the time. It only came tied to Fairlane V-8's and its ratios were 2.73, 2.07, 1.51 and 1:1, which were fairly wide compared to the big-engine cars of the era.

And topping things off was the first production high-performance small-block V-8 for the Fairlane, the Challenger 4-V/289 High Performance V-8, or 289 HP as it became known.

This was a genuine performance engine with nothing simulated. Bore grew to an even four inches, yielding the 289 inches. Stroke was still very short, meaning it had potential for high rpm. The advertised output of 271 horses at 6000 rpm took advantage of that characteristic. Torque of 312 pounds-feet came on at a healthy 3400 rpm. Compression was boosted to 11:1.

Other modifications included mechanical lifters, high-lift cam, special heads and a free-breathing four-barrel carburetor. On the other end of the operation were streamlined cast-iron exhaust headers. The hot 289 listed for $424.80 extra and was not available in wagons or with automatic. Road testers were now under nine seconds 0-60 and reporting quarter-mile times approaching sixteen seconds, with speeds nearing ninety.

Usually there is a drop-off in sales after the first year a new car goes on the market, but the 1963 Fairlane bucked the trend, with production rising 19.3 percent to 343,887 while the domestic industry built 9.5 percent more cars than in the 1962 run.

Once again the Fairlane 500 four-door was the most popular, but the hardtops did well with the Fairlane 500 version netting 41,641 orders and the Sports Coupe 28,268.

One 1963 Fairlane hardtop turned out to play a significant role in the immediate future of Fairlane performance cars. It appeared at the NHRA Nationals Labor Day weekend at Indianapolis Raceway Park. Entered by Providence, Rhode Island, Ford dealer Bob Tasca, it was almost unnoticed in the sea of hot stocks.

Tasca, one of the most performance-minded dealers in the country, had experimented running a 406 with a Lincoln automatic in a 1962 Fairlane in drag racing. The combo did well in gasser competition and the groundwork was laid for a more sophisticated machine.

With the cooperation of Ford's Special Vehicles Department, Tasca put a 427 in the Fairlane, with a fiberglass hood covering the high-riser intake manifold. Body panels were fiberglass, sound deadening was nonexistent and accessories were nil.

Since the mandatory fifty copies weren't made to qualify for S/S, Tasca had to run in the top factory experimental class (A/FX) where he turned a respectable 12.21/118.42. However, Dave Strickler topped the class in his lightweight 1963 Chevrolet Z-11.

Fairlane was not without some glory at the Nationals, thanks to Michigan driver Milo Coleman, who drove his 289 1963 Fairlane two-door sedan to D/stock honors, edging Galaxie driver Burl Kuhn. Coleman's class win was one of only two for Ford in the stock classes during the meet.

1964

Model year 1964 greeted the Fairlane with its stiffest competition yet. Four new intermediates came forth from General Motors: Chevrolet Chevelle, Pontiac Tempest, Oldsmobile F-85 and Buick Special. All had 115-inch wheelbases and outer dimensions in the Fairlane range. Underneath were conventional frames, similar to those on the big cars and up front was enough room to stuff the biggest engines GM made, even though there was a self-imposed corporate limit of 330 cubic inches. A wide variety of body styles was available, including some very good-looking convertibles.

Fairlane had one less competitor, Meteor, as Mercury decided to concentrate on its compact Comet line.

To make matters worse, Pontiac found a way to get around the GM engine rule by offering a GTO option for its LeMans series of the Tempest, which included the big Pontiac 389-cid engine with ratings up to 348 hp.

Chrysler made things even tougher with a one-two punch for its intermediate-sized cars. It started with a hotter wedge 426 for racing and offered a 365-hp street version. Then it unleashed its awesome Hemi for racing applications.

Ford, while deeply into its Total Performance program, couldn't do much to counter the tough GM and Chrysler intermediates on the street, as there just wasn't room under the hood of a Fairlane for a production big-block, without major changes. But that didn't mean a limited production car couldn't be built, as we'll see later.

Nineteen sixty-four styling was a facelift. Sedans got smoothed-out roofs and all models received rebent sheetmetal below the beltline from about the middle of the car back, resulting in the removal of fins. Taillights were enlarged, the rear quarters were bulged a bit and width went up 0.9 inch to 72.2 inches. A new cross-hatch grille graced the front and side trim was revised to be lower on the body panels.

Though billed in Ford literature as "the family-size total performance car," the 1964 Fairlane's most powerful punch was the 289 HP V-8 at 271 hp, which couldn't be had with an automatic or in a wagon.

"So O.K.; get in and show 'em displacement isn't everything!" said a Ford magazine ad for 1964. But it must have had a hard time convincing the folks, considering the competition.

There were a couple of other mechanical changes for the 1964 models. A 200-ci six that was added mid-way through the 1963 model run, was the only six to which you could hook an automatic. The 221 was buried and a mild version of the 289 was added, rated at 195 hp, thanks to a two-barrel carb and 9:1 squeeze. The new 289 could be had with a dual-range Cruise-O-Matic three-speed, the first such automatic for Fairlane.

Domestic new car production advanced just over 7.5 percent for 1964, but the Fairlane did not share in the wealth, as the increased competition left its mark with sales dipping 19.3 percent to 277,586, the lowest so far for it.

The Fairlane 500 four-door sedan preserved its top ranking with 86,919 orders. Combined, the two hardtop models accounted for 64,164 units, which was only off 8.2 percent from the year earlier.

While the hardtops may have been the sportiest, they weren't the fastest 1964 Fairlanes, that title belonged to a group of fifty-four Fairlane 500 two-door sedans called Thunderbolts.

Ford's stock car racing philosophy, be it in circle-track or drag competition, was to race its sales leader, which at the time was the big Ford Galaxie. While this worked well in NASCAR, USAC and other sanctioning groups there was a problem on the drag strips.

While there were no excuses to make for Ford's 427 big-block, which the engineers kept updating, it had to haul around the big Galaxie body and do it against the smaller MoPars, which also had strong engines, the 426 wedge.

In 1963, Plymouth, Dodge, Chevrolet and Pontiac all sold lightweight cars to drag racers. Ford did the same with a run of fifty lightweight Galaxie hardtops.

New sheetmetal eliminated the fins from the 1964 Fairlane. The Fairlane 500 Sports Coupe got newly designed spinner wheel covers and other identification. The V-emblems on the front fenders were similar to the Mustang, introduced during the model year. This one had the 260 V-8.

A GM edict to get out of racing left the Chevrolet and Pontiac lightweights under the required fifty units to be legal for the top stock class, S/S, but the MoPars made the grade and proceeded to eat the Fords alive, being close to the 3,200-pound limit for the class. They undercut the Galaxies by about 300 pounds in some cases.

Since power was close, the MoPars dominated their class. They were also helped by having an extremely strong automatic, Torqueflite, which shifted well, had a high stall speed and withstood the output of the big wedges. Ford didn't even sell street 427's with an automatic at the time, as the Cruise-O-Matic was felt to be not up to the task.

The light Chevys and Pontiacs spent the year in the factory experimental class, forcing Ford down to the lower classes if it wanted to win. Sales leadership aside, Ford had to do something; and as it turned out, the Tasca 1963 Fairlane was the tip of the iceberg. Maybe Ford couldn't shoehorn the 427 into a Fairlane on the production line, but in limited numbers, almost anything could be accomplished. Fairlane seemed the logical answer to the MoPars. It was a bit smaller and applying the lessons learned from its bigger lightweights, Ford had a potential S/S winner.

Andy Hotton's Dearborn Steel Tubing, which did special vehicle projects for Ford, turned out fifty-four 1964 Thunderbolts in all. Using the Fairlane 500 two-door sedan for a start, the first task was to get the 427 High Riser under the hood.

The high coil front suspension, with shock absorbers inside the coils produced shock towers that protruded inside the engine compartment and nearly extended to the hood. Since stock suspension design had to be retained, Ford could not go to coils between the front control arms, as on the bigger cars.

Instead, the upper A-arms were shortened and moved out, so the spring tower could be trimmed and strengthened. Special inner fender panels were constructed. Cowl-to-spring-tower bracing also had to be changed for clearance.

Even with those changes, the exhaust tubing manifolds had to be wrapped around the suspension components to clear. This was tricky, for design called for the four exhaust headers on each side of the engine to be the same length, thirty-one inches.

The engine was Ford's best at the time, the High Riser. It had two four-barrel Holleys on an aluminum high-rise intake manifold and high-rise heads with machined combustion chambers that fit on top of domed pistons and a 12.7:1 compression ratio. A 300-degree high-lift cam opened 2.20-inch intake valves and 1.73 exhausts. A forged-steel crank pushed up stronger rods, and looked down at a ten-quart oil pan.

Behind the exhaust headers were fourteen-inch-long collectors. A stock muffler and exhaust system came off one side of a crossover pipe, so the car could meet the rules. On the back of the engine in stick-shift models was a combination bell housing and scatter shield, cast in aluminum by R. C. Industries, which supplied similar units for the Galaxie lightweights.

A modified Galaxie driveshaft twisted a Detroit Locker differential. A couple of ratios were reported, with 4.56:1 the most common. Tough thirty-one-spline axles turned the rear wheels. Leaf springs were retained at the rear of the car, but they were only for supporting the weight, not for transmitting the power to the ground.

A separate system was set up for that chore. A square tubing cross-member was added ahead of the front of the rear leafs. It looped under the driveshaft, providing a required safety catch. The cross-member served as the forward anchor-

The car that turned around Ford's fortunes in super stock drag racing was the 1964 Fairlane Thunderbolt, based on the Fairlane 500 two-door sedan. The lightweight car had a 427 up front and greatly strengthened driveline and rear suspension. Gas Ronda is at the wheel, ready to pick up more points on the way to his NHRA Top Stock World Championship that year.

ing point for a pair of rectangular steel tubing traction bars that were welded and gusseted to the rear axle on either side.

Actually the leafs played a bit more of a part in the arrangement. There were two leafs on the left side and three on the right. A stiffer coil was used on the left front corner to preload the right rear spring. This was to help keep the right, rear tire on the ground, because the torque forces acted to lift it slightly and put more pressure on the left rear. The setup equalized that situation. Fifteen-inch steel wheels came with the car, but the racers tried all kinds of combinations to get the ideal weight transfer and traction.

This setup was the key to making use of the Thunderbolt's power. It had to be sorted out in early runs, but did the job. It was the most chassis engineering a major auto company did just for the purpose of drag racing stock cars up to that time.

Let's get up from under the Thunderbolt and look at the body. First, the High Riser had a problem, it wouldn't fit under the stock Fairlane hood, just like on Tasca's 1963. To cope, a fiberglass hood was designed to clear the air cleaner. At the back of the bubble for the air cleaner were a couple of screened openings. These were not to let air in (a system that was discovered in later years) but to let hot air out, so it wouldn't be trapped in the cramped confines of the Fairlane engine compartment.

The air cleaner was fed through a pair of six-inch flexible tubes that extended to either side of the cross-flow radiator and hooked up behind the inside headlights, which were removed from the bezels and replaced with screen.

Early Thunderbolts had sheetmetal air-cleaner housings, but the standard unit was a neat-looking cast-aluminum setup.

Hoods weren't the only fiberglass part of the Thunderbolt. Front fenders, doors and, on early models, the front bumper and splash pan were all of the light stuff—that saved two-thirds of the weight of conventional steel body panels. The early fiberglass front bumpers and pans raised objections by the drag racing groups. These were changed to an aluminum front bumper and brackets with a fiberglass pan.

Removing weight from the front allowed a more rearward weight bias for traction and also lowered the car's weight, approaching the 3,200-pound goal.

Windshields were still of safety glass, but side and rear glass were replaced by plexiglass. Rear quarter windows were fixed, with winding mechanisms omitted.

Inside were lightweight police bucket seats in front and the regular bench in the rear. Plain rubber floor mats sat on the bare floor, as nothing other than paint was applied. No sound deadeners or insulation were included. A Hurst shifter was added for stirring the four-speed models, and a tachometer was attached to the steering column to tell when to use it. An oil-pressure gauge was also added.

On the right-hand side in the trunk rode a heavy battery, to help the right, rear traction. At first Ford tried to use a 125-pound truck battery, but the sanctioning bodies frowned, so a ninety-five-pound bus battery replaced it, which still weighed twice as much as a stock one. When the package was done, it took gas in the tank to put the T-Bolt over the 3,200-pound limit. That's how close Ford played it.

In the standard-shift category, there were few problems, as the trusty Borg-Warner T-10 four-speed, or even Ford's new four-speed unit could be used.

When it came to the automatics, it was another story. To ignore the automatic category would be to concede half of the S/S competition to Chrysler, so Ford had to at least try.

At first, the Lincoln Turbo-Drive automatic was adapted. It was similar to Ford's Cruise-O-Matic three-speed, but had stronger innards. A special housing was used to bolt it to the 427, but its low 2100-rpm stall speed and sluggish shifts proved unsatisfactory. A torque converter from a 390 Cruise-O-Matic helped some, but it took a good part of the season to sort things out, and when it came down to Christmas tree time at the strips, the MoPars still had the best setup. However, development led to Ford offering its HX automatic for dragging, which resulted in moderate success.

Few realized at the time, but one of the neatest things was that you could order a Thunderbolt through your dealer, race ready for about $3,900. Considering what you got, it was stealing. Considering today's prices and new car performance it makes you want to cry.

Well, was all that effort worth it for Ford? Was it good enough to beat the MoPars? It certainly was!

Early Thunderbolt racing was done in A/FX until the first fifty units were certified, then it went to make a home for itself in S/S. The first major test for new equipment was, as usual, the NHRA Winternationals in February 1964 at Pomona. T-Bolts were legal for S/S and S/SA (automatic) battle and were ready to test the MoPars.

Among the strong T-Bolters were noted tuner Les Ritchey, strong runner Gas Ronda, Butch Leal driving for Mickey Thompson and all the way from Rhode Island, the Tasca Ford, driven by Bill Lawton.

The onslaught was a success: Only one MoPar, a 1964 Plymouth driven by Jerry Grosz, made it into the final three S/S rounds, and he fell on that one to Leal. The final round was a run-off between Leal and Ronda, who was running with his tired backup engine. Nevertheless Ronda was ready and won the class with a 12.05/120.16 effort.

Super stock automatic and stock eliminator were the same old story with MoPars prevailing; Fords only getting to the third-last round.

As the season grew, so did the length of the Thunderbolt S/S wins. Unfortunately, when stock eliminator time came around, the automatic MoPars could usually (but not always) beat the stick-shift T-Bolts.

Highlight of the year for the drag racing world is the NHRA Nationals at Indianapolis. Reports said that the S/S class had the closest competition in memory and when it was over, Leal's Thunderbolt was on top with an 11.76/122.78 run.

S/SA went to Jim Thornton's 1964 Dodge Ramcharger and the stock eliminator final rounds were embarrassingly free of Fords and full of Chrysler products. Roger Lindamood's *Color Me Gone II* 1964 Dodge took the title, as Hemis were king.

When the season ended, Ford was awarded the National Hot Rod Association Manufacturer's Cup, due to the efforts of its lightweight Fairlanes, Galaxies and Falcons. Ronda was the 1964 NHRA Top Stock point champion.

Fairlanes were not always winners in 1964 however. After controversy in NASCAR over the legality of Chrysler's Hemi in Grand National racing, a Fairlane stock car was tried. With less frontal area and the same power as the Galaxie, it was figured that the Fairlane might be successful against the MoPars, as in drag racing.

Tests with a 427 installed proved that the Fairlane just didn't handle, and besides, its 115½-inch wheelbase was a half-inch shy of NASCAR's 116-inch minimum. It would be a couple of seasons before the Fairlane would get the NASCAR job.

1965

Considering it was on the last year of its styling cycle, the 1965 Fairlane underwent an unusual number of changes, unfortunately not all for the better. It still was available only with small-blocks and sixes, due to the limitations of the front end design, but Ford now was heavily advertising the over-the-counter Cobra kits that could be added for greater performance, noting that setups up to 343 hp were available.

Even though they didn't involve big inches, there were several changes in the engine department. Gone were the 170 six and 260 V-8. A redesigned seven-

Larger exterior dimensions were one of the results of the restyling given the 1965 Fairlane. This Fairlane 500 Sports Coupe carried the slightly changed side trim and nameplates on the roof pillars and trunk lid. The 289 was the only V-8 that year and plaques for same were on the front fender, behind the wheel.

main-bearing 200-cid six was now standard. Three versions of the 289 were optional, starting with the two-barrel, regular-gas 200-hp version. New was a 225-hp, four-barrel job with a 10:1 squeeze and thirst for premium. And finally, at the top again, was the 289 HP at 271 horses, except in wagons.

Also departed was the two-speed Fordomatic. Three-speed Cruise-O-Matic was now available, in various forms, for all engines, including the 271-hp twister. Four-speed boxes could be had with the 225-hp 289 as an option, while they were required on the 271-hp job. Overdrive was optional only with the 200-hp engine. Standard for the six was the regular three-speed manual transmission, while the 200- and 225-hp powerplants came standard with the three-speed all-synchro unit.

Axle ratios ran from 2.80 to 3.5:1 for the regular engines, but the 271-hp came with 3.89 or 4.11 gears.

Added to the missing list were thirteen-inch wheels. Five-inch-wide, fourteen-inch wheels were now standard. Standard sedan tires were now 6.95x14 rayon jobs, while 7.35x14 nylon rubber was optional for sedans and hardtops and standard on wagons. Cars with the 289 HP also got the bigger tires in the package.

Wheelbase grew a half inch to 116 and was now at the NASCAR minimum, just in case. Rear tread also gained an inch and was now the same as the front at fifty-seven inches. That's not all that grew. New sheetmetal puffed the width up 1.6 inches to 73.8 and overall length gained over an inch to nudge the 199 mark.

Sheetmetal from the beltline to the rocker panel was changed in an effort to make the 1965 Fairlane look bigger. Rectangular moldings encased the quad, horizontal headlights with a new three-dimensional straight-bar grille between them. Fenders now jutted forward at the top. Side sculpturing was a somewhat confusing array of lines and chrome, while the squared-off trunk area ended with rectangular taillights, replacing the round units used for the previous three years.

Inside was a redesigned instrument panel. The Sports Coupe still came with a console and bucket-type seats standard, plus special exterior trim and deluxe wheel covers.

The model lineup remained the same, but the sales picture didn't. Production reached a new low for the Fairlane, dropping to 223,954, or 19.3 percent. Since the high point of 1963, Fairlane sales tumbled nearly thirty-five percent, or 120,000 cars.

Clearly the competition was taking its toll, and the lack of a performance image didn't help.

While the 1964 Thunderbolts may have been a hot topic in drag racing the year before, the 1965 models didn't follow it up. A change in NHRA Super Stock rules no doubt prevented a 1965 Thunderbolt from being produced in legal numbers. Minimum production for S/S was raised to one hundred units, double the

For 1965 were all-new designed optional wheel covers: full disc type (shown here) or wire-style type.

Fairlane accessories for 1965 included this padded instrument panel with luxurious cushioning topped with color-keyed vinyl.

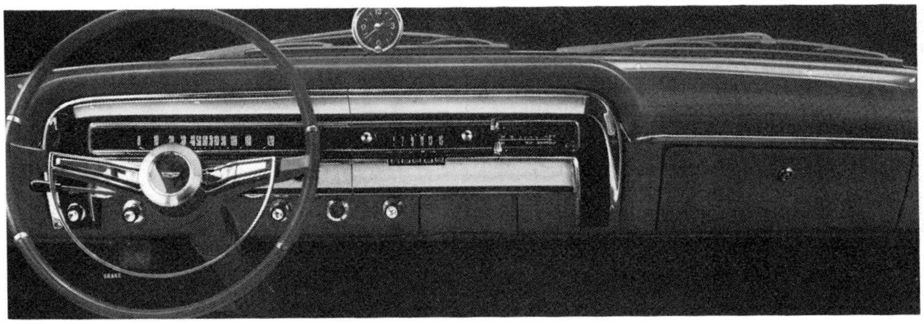

old requirement. Also lightweight panels were banned, as all had to be steel, and minimum weight was boosted to 3,400 pounds.

Factory experimental class rules remained the same as 1964, so Ford, which wanted to get its sohc 427 into action, went that route, using the smaller Mustangs and Falcons as starting points.

There was at least one 1965 Fairlane that came close to the Thunderbolts in configuration, however. Darrell Droke campaigned a lightweight Fairlane 500 two-door hardtop with an sohc aboard. He won his share of matches, including Mr. Stock Eliminator at the 1966 American Hot Rod Association (AHRA) winter drags at Irwindale, California.

1966

However, the job of establishing Ford in the performance segment of the intermediate new car market would rest with the redesigned 1966 Fairlanes.

Obviously, something was needed to catch up with the competition in style, performance and selection. Ford was well aware of the situation and countered with a completely redesigned Fairlane for 1966.

Showing rare restraint, the new package was close to the old one in dimensions, even shrinking in length, but it incorporated all the right basic ingredients to do the job.

First, the unitized construction and basic suspension designs were retained, but altered for the performance needs of the car. Front suspension continued to mount the coil springs above the upper A-arms, but the shock towers and pieces were redone so that big-blocks would fit better into the engine compartment.

Larger front springs were used. Rear leaf springs also were enlarged, being extended three inches to fifty-eight and widened to 2.5 inches from two.

Torque boxes were retained, but were no longer directly connected to the floor, but rather to other reinforcement to take noise away from the passenger compartment.

Body shells were truly all-new for 1966, with all styles being changed and convertible models being added for the first time. Sedans used a formal roof style, but in sleeker form than the previous models. Two-door hardtops had swept-back rear pillars that blended into the body to give a well-proportioned look. Fairlane hardtops and convertibles were among the best-looking intermediates of the year and a great contrast to the boxy 1965 models.

Vertical quad headlights, with the top lights jutting ahead of the lower, were at the outer ends of the front fenders, separated by a full-width, divided straight bar grille. Body lines were simple with a full-length ridge along the side, which was free of chrome on all models. There was also a lower crease segmented by the wheel wells.

Chrome trim, depending on model, was relegated to the rocker panels and wheel lips. There was a slight bulge for the rear quarter panels. Taillights were vertical rectangles, with trim on the bumper-height trunk varying with each model.

Inside was another new instrument panel, which was part of the body structure. Padding came standard. A panel extending about sixty percent of the way

Attractive all-new styling marked the 1966 Fairlanes. Among the best looking intermediates that year were the two-door hardtops, such as this Fairlane 500/XL example. Bucket-type seats were standard. The engine plaque on the front fender, behind the wheel, indicates a 289 V-8.

across from the driver's side contained a fuel gauge, horizontal-sweep speedometer and provision for a radio.

As changed as the cars, was the model lineup. The base Fairlane series came in two- and four-door sedans and a two-seat, four-door station wagon. Next up was the Fairlane 500 which had all of the above and added a two-door hardtop and convertible. New was the Fairlane 500/XL series in hardtop and convertible. At the top was Ford's answer to the muscle car sweepstakes, the Fairlane GT and GTA, which also came in hardtop and convertible models. There also was a Fairlane Squire wagon which, like other Fairlane wagons, offered a third seat as an option.

When it came to performance, the most interesting were the GT and GTA (called GT/A in some literature). First, both were the same basic cars and carried the same model body numbers, but differed in designation according to the transmission installed. Standard-shift cars were designated GT and those with automatics, which was Ford's new Sport Shift Cruise-O-Matic, which we'll get to shortly, got the GTA tag. Badges on the side and rear of the car changed accordingly.

Standard for GT's was Fairlane's first production engine from the FE big-block series, the venerable 390. The last hot 390 was the solid-lifter, high-performance mill in the early 1962 models, unfortunately the Fairlane's 390 didn't go that far.

Ford, like its competitors, had to provide performance for the masses, and that precluded hot cams, overly thirsty carburetors and valves that need periodic adjustment.

Ford did provide a higher-lift cam, redesigned cast-iron intake, larger four-barrel Holley carb and less restricted air cleaner to increase the horsepower reading over the other basic four-barrel 390's from 315 to 335 hp; but the number was achieved at an easy 4600 rpm. Compression ratio was 10.5:1. In 315-hp form, the 390 was an option on other Fairlane models.

Standard gearbox for the GT was Ford's three-speed all-synchro unit, with the Ford four-speed transmission optional. Four-speeds in 390 cars were close-

Using a Fairlane GT as a basis, Ford created the Fairlane GT A Go Go for displays at auto shows around the country. Stripes down the middle, special side trim with exhaust outlets and special vents in the hood were among the features. Noted customizer Gene Winfield had some input into the design.

Mercury switched its Comet name from a line of compact-class cars to the intermediate field for the 1966 model year. Like Fairlane, there were performance models, like this Cyclone GT two-door hardtop, which had buckets inside and a 335-hp 390 V-8 standard.

ratio with 2.32, 1.69, 1.29 and 1:1 cogs. This compared to the wide-ratio gears for 289 cars at 2.78, 1.93, 1.36 and 1:1.

Highly touted was the Sport Shift option which permitted manual or automatic shifting. Ford redesigned the valving in its Cruise-O-Matic three-speed for the chore. The C-6 transmission came with a floor linkage and was only to be available in the GTA's. Cars equipped with the 390's got nine-inch ring gears in the differential. Ratios ranged from 2.80 (for small engines) to 3.5:1.

Outside identification for the GT's consisted of an emblem in the center of the blacked-out grille, chromed indentations in the hood with the engine size in red numerals, GT or GTA plaques on the front fender, in line with vinyl side stripes which were parallel to and just above the rocker panel and a rear chrome trim panel between the taillights with a GT or GTA badge. At the end of the rear fender, the word Fairlane was spelled out in individual chrome letters, while other models had letters attached with a chrome line under them, as on the larger Fords.

Inside, bucket seats were standard, with a console between them. The console occupied the area from just ahead of the shift lever back, carved out on the driver's side and covered with a satin-finish silver panel. A GT badge was at the front of the console, beneath the dash. Doors also carried GT identification.

To top it off, full wheel covers were standard and the engine came with a chromed dress-up kit.

Sharing the bucket-seat-and-console interior with the GT's was the lower priced Fairlane 500/XL series. A six-cylinder engine was standard.

When 1966 Fairlane GT series cars came with Sport Shift automatic, they were designated GTA models. The redesigned version of the Cruise-O-Matic let the driver shift gears manually, or did it automatically. The console featured a storage compartment next to the seats and GT badge in front of the shift lever.

After waiting too many years, Fairlane got a performance series for 1966, the GT/GTA. A 390 V-8, bucket-seat-equipped interior and exterior ornamentation were part of the package. This GTA convertible shows off the side stripes, chrome hood decoration, blacked-out grille and grille badge the series featured.

XL models carried their own interior and exterior identification and side trim consisted of wheel lip, rocker panel and lower rear quarter panel bright metal. Full wheel covers were also standard. Both the XL and GT models featured red and white lights built into the door armrests. Rocker panel trim, lack of trim between the taillights and bench seats marked the Fairlane 500 models.

Dimensions of the 1966 models came out close to their predecessors. Wheelbase remained at 116 inches for all but wagons, which shared a 113-inch stretch with Falcon wagons. Overall length came down about two inches from the 1965's at 197, while width grew 0.2 inch to seventy-four. Wagons varied slightly in all directions.

Engine selection started with the 200-cid, 120-hp six for all models except the GT's. Options for non-GT models included the 200-hp 289, a 265-hp, two-barrel regular-gas 390 (275-hp with Cruise-O-Matic) and a four-barrel 315-hp 390.

Three-speed transmissions were standard across the board with V-8's getting all-synchro. Overdrive only came bolted to a 289 and four-speeds could be had with V-8's, providing they weren't surrounded with a station wagon body.

Tire sizes, on fourteen-inch wheels, ranged from 6.95 to 7.75. Brakes started the year with ten-inch-drum units standard front and back, though discs may have been installed on the front of some later models. They were already offered in full-sized Fords and Mustangs at the time.

The nice thing about the improvements to the 1966 Fairlanes was that they did not result in weight increases for the basic car. For example, the shipping weight on a Fairlane 500 two-door hardtop was 2,863 in 1965 and 2,856 in 1966.

GT and GTA for 1966 meant your Fairlane came standard with the new and exciting 390-ci four-barrel V-8, the Thunderbird Special.

New for 1966 Fairlane was instrument panel that conveniently grouped instruments, optional radio and SelectAire Conditioner controls for easy viewing and operation.

However, when the 390 was taken on board, the scales jumped, as more than 450 pounds were added. A GT two-door hardtop was listed at 3,324 pounds.

As if Fairlane didn't already have enough competition, it picked up another intermediate from within the company. Mercury brought out its 1966 Comet nameplate on the same chassis as Fairlane, complete with the 390 option and Cyclone GT performance models.

While it let its first intermediate, Meteor, die on the vine, the new Comet was heavily promoted in drag racing, an area where the Fairlane would be weak.

However, it wasn't the Comet the Fairlane GT was aimed at. Advertising pointed the volley in the direction of the car that ignited the hot intermediate fire, the Pontiac Tempest GTO. The GT engine was within a cube or so of the GTO's 389-cid engine and the Ford rating of 335 hp matched it exactly, even though the GTO hit it at 5000 rpm, 400 higher than Ford. Problem was that the GTO only *started* at 335. Triple two-barrel carbs were carded at 360 at 5200, which could be boosted by a dealer-installed fresh-air intake system.

One Fairlane GT ad gave a recipe, "How to Cook a Tiger." A toy tiger tail stuck out from under the hood. Unfortunately for Ford, the comparison backfired, as the GTO's did the cooking. *Car Life* tested a GTA and GTO, both in 335-hp trim, and found the GTO a full two seconds faster in the quarter-mile.

Actually, the GTO was just one of the mean machines in the 1966 intermediate muscle car battle. Chevrolet's new Chevelle SS-396 was readily available and offered an engine of the same size as the name with ratings that reached 375 hp during the model year. The Oldsmobile F-85 4-4-2 came standard with a 350-hp 400-cid powerplant and added a mid-year 360-hp option. Buick offered its Skylark GS with a 401-cid V-8 and ratings of 325 and later 340 hp. Chrysler out-gunned them all with its newly released street Hemi option for its Plymouth Belvedere, Dodge Coronet and Charger, rated at 425 hp.

About the same time the Fairlane GT's hit the drag strips, the word hit the streets that the 390's were running out of steam well before they reached the 5000-rpm mark, and that the other cars in their class, C/stock and C/stock automatic were feasting on them.

The answer wasn't far away, for the 390 and 427 shared the same basic block and the 427 had all the gingerbread for drag racing.

A limited number of 1966 Fairlane 427's were built, about seventy. Super stock drag class eligibility was dropped back to fifty, from one hundred for the 1965 models.

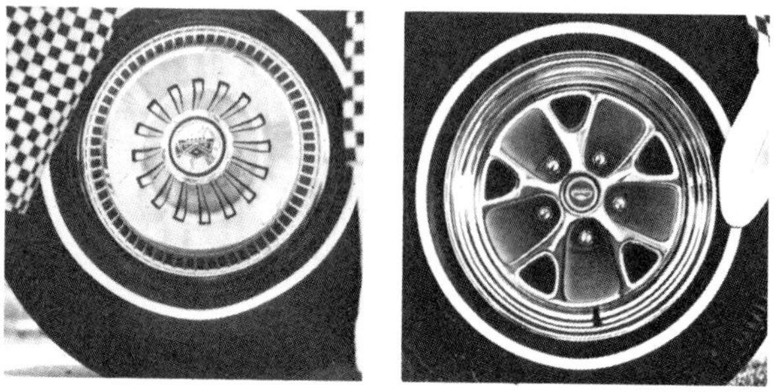

The option list for 1966 included: left, bright-finished deluxe wheel covers and, right, sporty styled steel wheels.

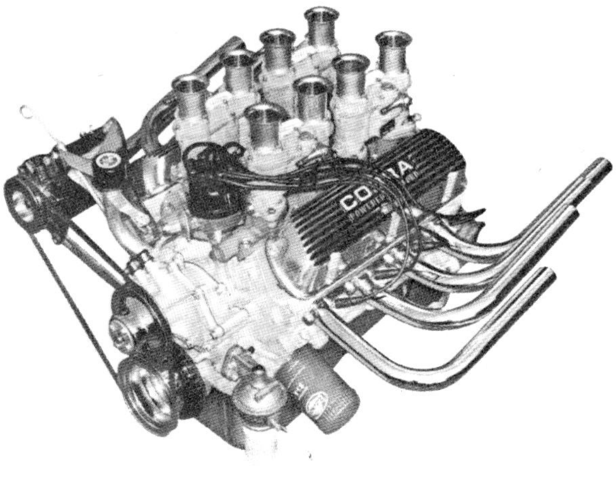

Available for the 1966 Fairlane were Cobra kits with special exhaust header pipes, shown here on the 289 V-8.

Both 410-hp single four-barrel- and 425-hp twin four-barrel-carb engines were listed, but most, if not all, of the Fairlane 427's were the latter. Also, the standard setup was the Ford top-loader four-speed. To clear the streamlined cast-iron exhaust headers, shock tower support plates were machined for clearance.

Installation was allowed in Fairlane and Fairlane 500 two-door sedans, Fairlane 500 and 500/XL hardtops and possibly some convertibles. (Note that GT's did not get 427's.)

Most visible evidence of a 1966 Fairlane with a 427 was the fiberglass hood with built-in scoop at the leading edge, which channeled fresh air right to the air cleaner. The hood was without hinges and was attached with four hood pins. When the pins were removed, the hood was simply lifted off.

The 1966 drag racing 427 Fairlane was not very successful at first. NHRA put it in A/Stock, where it would compete with the more powerful street Hemi. If that weren't bad enough, when it came to the National Championship drags at Indy on Labor Day weekend, there were more problems.

NHRA rules for stock classes said fiberglass body panels, hoods, etc. could not be used, unless a minimum of 500 cars were produced with them. Since only about seventy Fairlane 427's were made, they were put into the top factory experimental class, A/FX, where they would compete with things like sohc Mustangs, which were entirely different types of cars.

The MoPars dominated A/S and the Fairlane 427 drivers sat it out. In later years, when classes changed, the Fairlane 427 became a very successful drag race vehicle.

Mercury Comets were active in drag racing with factory-backed efforts in the match racing class. Match race cars evolved out of the hotter FX cars. By the 1966 season, fiberglass bodies that resembled stock cars, but that were put over dragster-like chassis were running in match racing events. They would soon be called "funny cars" and they grabbed the spotlight away from the hotter stocks.

Early participants included Comets, Ford Mustangs and a variety of Chrysler and GM products.

Fairlanes were by no means without racing success in 1966, however, thanks to a change in the racing climate in NASCAR. Through the 1965 season, the full-sized Galaxie was the only way to go.

With upper limits on the cubic inches of engines, other ways to increase speeds were sought, the smaller car being one of them. Galaxies had 119-inch wheelbases and were nearly seventy-nine inches wide. Fairlanes for 1966 were five-inches narrower and had three inches less wheelbase. While weights were regulated by engine displacement, the size of car and frontal area that could be pushed through the air could be reduced by going to a Fairlane.

Ford's philosophy was to race its most popular car, the Galaxie, but as the 1966 season progressed, that urge was superseded by Ford's desire to win; and a smaller car was the answer.

Comet styling for 1966, like Fairlane, was smooth and without frills. Showing off its lines is this Caliente convertible.

Actually a lot of pioneering effort with the new Ford Motor Company intermediate was done by independent Bud Moore, who fielded and honed a very successful 1966 Mercury Comet driven by Darel Dieringer. Culmination of that enterprise was Dieringer's win that year in the Labor Day Southern 500 at Darlington.

Early season factory Ford efforts were still in 1966 Galaxies for NASCAR Grand National Competition, but Curtis Turner did experiment with a Fairlane in short-track events and took a second in the April 3 hundred-miler on the dirt oval at Hickory, North Carolina.

A little less than two weeks later, Ford withdrew its factory support of teams from NASCAR, over a dispute regarding legalization of the sohc 427.

Factory Fords were parked, but Moore's Comet was not and when Ford decided to return to NASCAR, the Galaxie days were numbered. Late in the season, factory Fairlanes were running on equal terms with Plymouth and Dodge Hemis. Not all the credit can go to the switch to the Fairlane, however, as a rule change permitted Ford's wedge engines to have two four-barrels to one for the Hemis.

Fred Lorenzen gave the Fairlane its first major win on September 25, 1966, when he wheeled the Holman & Moody number 28 home first in the 500-lapper on the Martinsville, Virginia, half-mile paved oval. On October 30, Lorenzen ended the season by winning the American 500 at Rockingham, North Carolina. With the exception of second-place Dodge driver Don White, Fords filled the top five.

Not only did the 1966 Fairlane end up a winner on the racetrack, it played the same role in the showrooms. Production ended with a whopping 41.7 percent increase, compared with an industry average of only 9.2 percent. Orders pulled 317,274 Fairlanes from the production lines, second only to 1963.

Underlining the success of the hardtop styling, the Fairlane 500 model was the most popular at 75,947, edging its four-door-sedan-series mate by 7,312. Surprisingly, GT's outpaced the XL's with 37,343 being built to 28,502. Convertibles, despite their good looks, remained relatively rare with 9,299 Fairlane 500's, 4,560 Fairlane 500/XL's and 4,327 GT copies being made. Comet production for the 1966 models was more than the 1962 and 1963 Meteor combined, at 170,426.

1967

For Ford and just about all the other competitors in the intermediate field, the 1967 model year was one of refinement and not major changes. Only the American Motors intermediate cars had major changes, and with 343 cubic inches the largest powerplant, they hardly had an effect on any performance doings.

Fairlane went with only minor changes in style and mechanics. A new grille pattern divided a smaller mesh screening into eight segments, while new split taillights filled the old housings. Side trim consisted of a full-length wide strip of chrome

A new grille design let people know this was a 1967 Comet. This Cyclone two-door hardtop wears simulated wire wheel covers.

and anodized aluminum from front bumper to back, about a quarter of the way up the body. GT's retained their stripes and lower-priced Fairlanes got a chrome spear just below the side crease, running from the taillight to the front wheel well. Models that got trim between the taillights on the trunk also got a narrow band of bright metal.

Model lineups remained the same for passenger cars, but the Ranchero pick-up, which was based on the Falcon from 1960 through the 1966 model years, was now moved into the Fairlane category, via a Fairlane front clip. Sharing the 113-inch wheelbase with wagons, it came in Fairlane, Fairlane 500 and Fairlane 500/XL designations. It was counted with truck sales and production.

Back for its second year at the top of the Fairlane passenger car totem pole was the GT series. There were a number of equipment changes, but most notable was the replacement of the 390 as the standard engine with the regular-gas 200-hp 289.

While the horsepower may have gone down, the stopping power went up for the 1967 GT models. Power front disc brakes were now standard for them and optional for other models.

The decrease in engine size and addition of the discs resulted in relatively modest price increases for GT models for 1967, compared to the rest of the line. For example, a GT convertible listed at $3,045.76 for 1966 and $3,063.67 for 1967. This compared with a 500/XL convertible listing of $2,747.92 for 1966 and $2,843.32 the following year.

GT identification changed somewhat for 1967. A crest replaced the round grille identification, chrome GT letters replaced the side badges on GT models and a small GTA plaque was used for that variation. Rear plaques were also crest shaped. The chrome hood trim was replaced by twin "power domes" that incor-

After being closely associated with the compact Falcon since the 1960 model year, Ranchero became a Fairlane for 1967 getting a front clip from the intermediate and Fairlane trim. This 500/XL was the top of the line that year.

Heavier side trim distinguished the 1967 Fairlane 500 and 500/XL models. This XL hardtop carried its series badge on the rear fender and an emblem on the right side of the trunk lid. Taillights used the same housing as '66 but were redone for 1967.

porated turn-signal indicators. When equipped with a 390, the chrome numerals were attached to the rear of the outside of the domes.

Engine changes were concentrated on the 390. Instead of four versions, there were now two, a two-barrel unit at 270 hp and a four-barrel at 320 hp. For GTO and other such hunting, Ford now openly advertised its 427, in two versions, 410 and 425 hp, though it is doubted that many 410's were made, if any.

The fiberglass hood with scoop was an option for 427 Fairlanes instead of being standard as was the case the year before. Availability of the 427 continued in the selected models from 1966 and again did not include the GT series. As in its other lines, Ford hung the Cobra name on its 427 engines for 1967, to cash in on the famous Shelby sports car. It would be attaching the Cobra tag on many items for years to come.

New car production dropped nearly eleven percent for the 1967 model year, but Fairlane was especially hard hit, losing most of what it gained the year before. Production fell 24.8 percent to 238,688, less than 15,000 above the 1965 low. Remaining the most popular model was the Fairlane 500 hardtop.

Despite the moderate price increase, GT series sales fell 44.3 percent to 20,787. While the drop was severe, it wasn't as bad as the Mercury Comet, which received more of a facelift than the Fairlane for 1967. Mercury launched the Cougar sporty compact and didn't pay a whole lot of attention to the Comet in promotion, and the customers followed suit. Production fell from 170,426 to 81,113 or 52.4 percent.

In late model stock car racing, the 1967 Fairlane also was not quite the success it should have been.

Looking at NASCAR, first, the fun and games with the rules continued. First, both Chrysler and Ford announced late in the 1966 season that they would be trimming their stock racing budgets for 1967.

NASCAR also said it would finally OK the sohc 427 for use in Fords with no weight penalty and rumors spread of a production model Galaxie with the sohc installed.

At roughly the same time, Congress was in the process of setting up the machinery to regulate the auto industry and taking a dim view of performance promotion at the same time. Ford had tried since 1964 to get the sohc legalized in a fair manner, but after finally having won the battle, it decided not to use it in the performance war, for fear of the negative publicity of offering an engine capable of over 600 hp in a production car. The 427 wedge would have to do for another season.

Even though Ford did not take advantage of the sohc rule, builders of 1967 Fairlanes did take advantage of another regulation change. The Grand National rules for 1967 no longer called for a completely stock chassis. That permitted Fairlane constructors to do away with the high coil springs and shock towers and substitute the front frame stub from the 1965-66 Galaxies, which had coil springs be-

Back again for 1967 was the attractive Fairlane two-door hardtop, here in GTA form. Carryover features of the line were blacked-out grille (with a new badge and pattern), side stripes and GT identification. This example has the optional styled steel wheels.

tween the control arms. This created more room in the engine compartment, a stronger front suspension (the 1965-66 Galaxie front stub is the strongest in the business and is still used for Grand National stock cars today), and being a known factor, helped the handling of the smaller Fairlanes.

At the end of the 1966 season, Ned Jarrett retired from the Bondy Long Ford team. He was replaced by Dick Hutcherson. Junior Johnson also retired from driv-

Still at the helm of the 1967 Mercury Comet lineup for performance fans was the Cyclone GT. This two-door hardtop had the fiberglass hood with simulated scoops and wheel covers that looked like chrome wheels. Striping was similar to Fairlane GT's.

The 1967 Fairlane 500/XL convertible interiors were luxurious; with full carpeting, pleated vinyl upholstery, deep-foam bucket seats and deluxe seat belts.

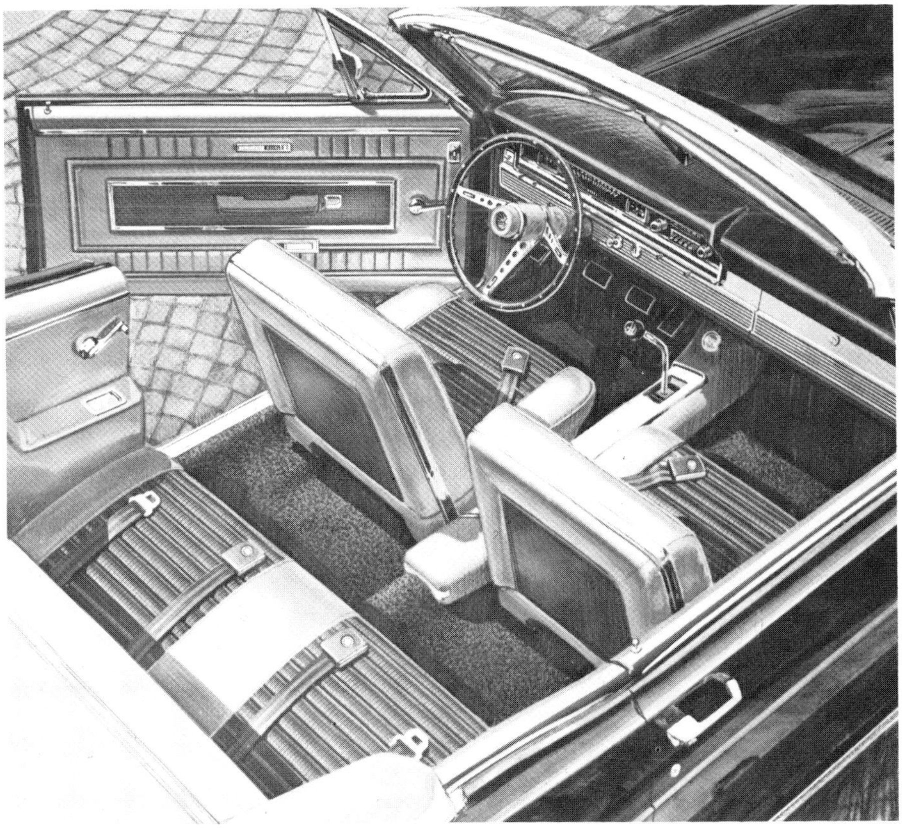

ing after 1966, but continued to enter Fords, settling on Darel Dieringer (who formerly drove for Bud Moore) as his driver.

The Wood Brothers continued with Cale Yarborough, whom they picked up the year before. Fred Lorenzen continued to hold down the seat in the first Holman & Moody car.

First date on the 1967 NASCAR calendar (one 1967 season event was held in late 1966) was the fifth annual Motor Trend 500 at Riverside, California. Dan Gurney won the first four, all in Fords and at least one string would be broken, as Gurney was entered in a Bud Moore Mercury Comet.

Hutcherson made his new owner happy by putting his Fairlane on the pole. After 135 miles, rain halted the race and forced a one-week postponement from

Former Dodge driver David Pearson won two straight championships for the Holman & Moody Ford team.

After retiring as a driver, Junior Johnson played the role of a car owner in 1967.

Four 1967 Fairlanes make their way through the north turn at the Wisconsin State Fair Park Speedway on August 17, 1967. After wrecking his own car in practice, Parnelli Jones (#121) drove for Jack Bowsher and won the race. Following are Mario Andretti (#11), A. J. Foyt (#27) and Bowsher (#21).

January 22 to 29. When the race resumed, Gurney ran into all kinds of problems and eventually left the track with a broken rod.

Parnelli Jones, after years of being the second-fastest driver to Gurney, went on to victory in a Fairlane built by Holman & Moody, prepared by Bill Stroppe and crewed by The Wood Brothers. Despite the announced Chrysler cutbacks, two Plymouths and two Dodges followed Jones across the finish line.

The win gave Ford a chance to advertise its 427 Fairlane. The ad showed Jones's winning car with the number 115 on the side. The caption read, "The 427 Fairlane is also available without numbers." A stock Fairlane 500 hardtop was below, complete with an incorrect 427 plaque on the front fender and stock hood.

More good Ford news would come from Daytona. In 1963 and 1964, the competition surprised Ford racers with something new, now it was Ford's turn. New tunnel-port heads and intake manifolds were ready for the 427.

Like the Chevy 427 in 1963 and the Chrysler Hemi in 1964, the new equipment for Ford had not been made in sufficient numbers to meet the rules, but NASCAR looked the other way.

The heads utilized large, round intake passages, made possible by a new tunnel-port design. Previously, intake passage size was limited by being routed between the pushrods. On FE series big-blocks, the pushrods went through the outer edge of the intake manifold. On the new setup the runners went through the pushrod area with separate tubes surrounding the pushrods right in the middle of the passage.

The single-plane long intake manifold for the two four-barrel carbs Ford was allowed to run, also had large round runners to match the heads. The result was a very free-breathing 427 engine.

Chrysler, struggling with a reduced budget and basically last year's engineering, was furious. Their blood pressure wasn't helped one bit to hear the NASCAR explanation of why the Fords were allowed to run: If NASCAR made the Ford teams switch back to the old heads, the new intake manifolds wouldn't fit.

Tunnel-port heads and intake were allowed. Ford would be using the concept in future engines, but not for the production 1967 Fairlane.

Once the Fords got sorted out, defending USAC Indy-car champion Mario Andretti was turning laps in the 182 range in his Holman & Moody Fairlane.

The hundred-mile preliminary contests had mixed results with Dodge Charger driver LeeRoy Yarbrough taking the first and Lorenzen the second. However, the 500 turned out to be a Ford contest with Andretti first and Lorenzen second, giving Holman & Moody a one-two finish.

The next superspeedway contest came up on April 2, the Atlanta 500. Cale Yarborough gave Ford its third major win of the young season in The Wood Brothers car and Hutcherson backed it up with a second-place finish. Chrysler threatened to withdraw altogether and even tried a boycott of the race.

However, after that Ford's fortunes changed and Richard Petty went on to the winningest season ever for a NASCAR Grand National driver. Piloting Plymouths, Petty scored twenty-seven victories during the season, including another record, ten in a row. Petty easily won the title and Plymouth's thirty-one flags topped all manufacturers. Ford repeated its 1966 total of ten wins, which is even less impressive when you consider that factory Ford teams sat out a good part of the previous season.

Mario Andretti won the 1967 Daytona 500 in a 1967 Fairlane hardtop owned by Holman & Moody and powered by a 427 with new tunnel port heads which made the Fairlanes the fastest cars on the track.

During the Atlanta 500, Lorenzen pulled out after a blown engine covered his car with oil. It turned out to mark his retirement as he didn't race for Holman & Moody the rest of the year, citing health and other problems.

Lorenzen did stay with the organization, helping other Ford competitors from the pits. Late in the season he headed a team to put impressive youngster Bobby Allison in a Fairlane. Allison, who raced his own cars, responded by winning the last two races of the season, the October 29 National 500 at Rockingham and the November 5 250-miler at Weaverville, North Carolina.

Holman & Moody's replacement for Lorenzen turned out to be a good one. David Pearson, who won the 1966 driving championship in a Cotton Owens Dodge, had just split with the team and agreed to drive the Fairlane. He didn't win a race for them during 1967, but better days were coming.

Ford hadn't had a strong USAC stock car effort since it backed the Zecol team in 1964, but for 1967 all stops were pulled out to take the northern circuit by storm.

For the bigger races, especially at Milwaukee, Ford had Jones in a Holman & Moody/Stroppe car, Andretti in a Holman & Moody Fairlane like he won with at Daytona, A. J. Foyt in a Banjo Matthews Fairlane and Jack Bowsher in his own cars. Independent Whitey Gerken also had a little help with his new Fairlane.

Bowsher's operation was interesting, as he fielded an independent Galaxie in 1966 and was generally the fastest of the Ford drivers. In 1967 he showed up at the tracks with fully equipped vans and a helper on the crew by the name of Lorenzen.

Ford drivers took nine of the twenty-three events on the USAC schedule, but together they could only equal the performance by Don White, who also scored nine victories in a Ray Nichels Dodge Charger and won the championship.

Parnelli Jones won three races—he only started a dozen events compared to White's twenty-two—and finished a fairly close second in points. One of Jones's victories came in a Bowsher Ford, when his own ride was wrecked in practice. Bowsher scored four wins and one each went to Andretti and Foyt, both coming in the same car, as Foyt abandoned the Matthews Fairlane late in the season.

In drag racing, the stock rules changed again. Trying to figure out where your car would be placed each year was like putting an X on the side of a boat to mark a good fishing spot in the lake.

When last we left the NHRA stock classes, S/S was the hottest for cars which made production quotas, then lesser vehicles went into A/S, B/S, C/S and down.

For 1967 there were now five levels of super stock competition, starting with SS/A and going down to SS/E, depending on the weight and power combinations. Cam design and tire sizes were open.

Fairlane 427's fell into SS/B. Even though there was strong competition from the MoPars, Fairlanes had their shining moments, including Ed Terry's class win in the Nationals aboard a 1966 Fairlane 427.

1968

Every once in awhile, a model year comes along where many makes and models are changed at the same time. That's what happened for 1968, although it wasn't exactly a coincidence as a large number of emission and safety regulations took effect for the first time, calling for many changes.

Mercury's main performance promotion for its Comets was in drag racing, where fiberglass bodies were made to resemble a '67 Cyclone hardtop. Jack Chrisman checks the supercharged, overhead-cam engine in his funny car. The chassis was strictly for racing and the only similarity to the stock car, in this case, was the wheelbase.

In the intermediate field, exactly the opposite of 1967 took place as every one of the entrants was completely restyled, except those from American Motors. Competition would be keen, both in terms of performance and looks.

General Motors intermediates for 1968 were the most changed with all-new styling and the use of two different wheelbases. Two-doors were on a 112-inch frame and four-doors, 116. Both were on a 115-inch wheelbase in the past, except for some larger glass-topped wagons, which were on a 120-inch wheelbase both years. Each line had its hot performers: the Chevrolet Chevelle SS-396, Pontiac Tempest GTO, Oldsmobile F-85 Cutlass 4-4-2 and Buick Skylark GS 400.

Chrysler and Ford intermediates restyled the bodies, but the chassis were carryovers.

Plymouth brought out its traditional top-line performance series, the GTX, and added a new twist: a basic two-door sedan with a warmed-up 383, four-speed and heavy-duty suspension standard. Also added were a cheap horn and a neat gimmick, calling it the Road Runner. It was promoted, sold like crazy and spawned a whole field of econo-muscle car imitators.

Dodge restyled its Charger with a sexy new set of curvy sheetmetal. Its standard intermediate Coronet also was redone. Both got top of the line performance models dubbed R/T. Mid-year a basic two-door sedan was added to the Coronet line, the Super Bee, which directly copied the Road Runner.

With all that to go against, what did Ford do? Turned out what some consider to be the best looking Fairlane of all time, that's what. Using its basic 1967 under-

Two styles of two-door hardtops were in the 1968 Fairlane line: the formal roof job, shown here in Torino series form, and a fastback. Performance fans, race drivers and customers perferred the fastback.

Looking almost like a full-sized car is this 1968 Fairlane 500 convertible with the optional wheel covers. Side vent windows were eliminated from hardtops and convertibles that year. In front of the front fender is the 390 V-8 plaque.

pinnings, new bodies were designed for all but the wagons and Ranchero; and even for them, all tin below the beltline was new.

A growing process started that would take Ford intermediates close to the size of the big Fords. Length for non-wagons hit 201.1, up 4.1 inches. Width picked up a half-inch to 74.5. Height for the two-door hardtops dropped from 54.5 to 53.5 inches. Weights increased an average of 120 pounds.

Names also were changed. Top-line models now used the Torino designation, named after the Italian city where the major auto stylists are located. At the top of the line was the sporty Torino GT, which came in formal roof and fastback two-door hardtops and convertible. Under that was the Torino, which was aimed at the luxury buyer. It was offered in four-door sedan, formal two-door hardtop and Squire station wagon.

The Torino name turned out to be so popular that some thought it was the name for the intermediate Fords for 1968. Actually the line was still called the Fairlane, but now models with that name occupied the bottom two rungs on the ladder.

Lowest-priced intermediate Fords for 1968 were the Fairlanes, in four-door sedan, two-door hardtop and four-door wagon form. The formal-roofed two-door hardtop was the lowest-priced model, as two-door sedans were dropped from the 1968 range.

Next up was the Fairlane 500, which had the most models, a four-door sedan, both formal and fastback two-door hardtops, a convertible and four-door wagon.

Easily the most strikingly styled Fairlane was the fastback. It looked like a blown-up 1967 Mustang fastback with reverse-angled rear pillars. Ford stylists

Ford's Design Center came up with the Torino Machete for the shows in 1968. Based on the GT fastback, its highlights included a new grille with hidden headlights, no door handles, high-back bucket seats and hidden taillights. The styled steel wheels were available with plain hubcaps for non-GT models.

Mercury also had access to the Fairlane fastback design and did an excellent job with its 1968 Montego Cyclone, which had a different grille, taillight setup and trim details. This example sports wire wheel covers and an emblem for the 302 V-8.

seemed to pull it off just right, with a huge flat rear window, tasteful lack of chrome, especially in Torino GT form, with complimentary striping.

Making the car larger, stealing lines from a smaller edition and using an older chassis sound like a formula for failure, but it was just the opposite with the 1968 Fairlane. Sales boomed, racers loved aerodynamics and the press raved. Ford had a winner.

Looking closer at the styling, a rectangular grille housed the quad headlights, which returned to a horizontal format. Mesh sections were split into upper and lower halves with Torinos getting a center badge. Fenders projected forward of the grille like blinders, in which the parking lights/running lights/turn signals were mounted. Twin creases headed to the back of the car similar to the earlier Fairlanes. Rear quarter panels kicked-up and on fastbacks kind of blended into the sloping roof.

Rectangular taillights came in different designs for fastbacks, wagons and other models. Fastbacks used the widest taillight setups with three-lens sandwiches, red top and bottom with white for backup lights in the middle. Trim between the lights varied per model. Sedans and formal hardtops had trunks which opened to the bumper, while the small fastback lids only opened to the top of the panel between the taillights.

Formal hardtops had thick and gentle sloping rear roof pillars. All hardtops and convertibles did away with the front vent windows, while sedans and wagons retained them. A tinted rear window was standard on the fastbacks, as the sun could really bake the rear seat passengers.

Fairlane 500 models had a simple chrome strip down the side and between the taillights. Bench seats were standard with bucket seats and a console optional on two-doors.

Torinos got upper and lower chrome strips and fancier bench-seat-equipped interiors. For sportiness, Fairlane buyers had to go one step up.

A new feature of the 1968 Fairlane was a four-dial dashboard. What was in the dials depended on how you hit the option list. This Torino GT dash had the optional tachometer and clock in the right two coves. It also had a console and four-speed transmission. Consoles were standard at the start of the model year and optional later, on GT models.

Strikingly styled for 1968 was the Fairlane fastback two-door hardtop, shown in Torino GT form. It was a winner in sales and on the racetrack. Visible are the standard C-stripe, styled steel wheels with GT hubcaps, white band tires and wheel lip moldings. This early 1968 model had a 302 V-8 engine badge on the front fender.

Torino GT models started out with more standard equipment than they ended with. At the beginning of the model year, bucket seats and a console were standard, but in December of 1967 they were made optional with bench seats, which were formerly optional, now standard. Also Ford's new small-block 302-ci V-8 was standard at the start, but it joined the option list in December when the 289 V-8 replaced it as the standard powerplant.

Standard equipment that didn't change on the GT models included styled steel wheels with wheel rim rings and small hubcaps, complete with GT symbol in the middle, GT emblems front and back and tape stripes. Formal hardtops and convertibles got stripes above the rocker panels, which continued on the rear fenders. Fastbacks got a tapered set that swept along the top of the fender line, followed the leading edge of the front fender and swept back to the rear wheel opening.

Inside, Fairlane interiors were also redone. The dashboard was completely new with four big, round dials deeply recessed into the rather plain, padded panel. Optional tachometer and clock could make them all full, but without them, only three had gauges in them. The speedometer was second from left.

Console-equipped interiors got a newly styled unit that was attached to the bottom of the dash. Styling was more conventional than the good-looking 1966-67 consoles.

The Ranchero received the new skin below the beltline, but wasn't as closely tied with the Fairlane name. Models were Ranchero, Ranchero 500 and bucketed Ranchero GT.

Engines also underwent changes for 1968. Not only did their performance and economy have to be taken into account, but also their ability to meet the new federal emission standards.

Starting from the bottom, the 200-cid six was standard for all but the GT models and had its rating cut to 115 hp from 120. Standard on GT's and optional on other models at the start of the model year was a new number, the 302, achieved by increasing the stroke to an even three inches from the 2.87 figure that had been used for the 221, 260 and 289. The 289's four-inch bore remained.

The 302 had significance in that it was just under the five-liter limit for SCCA (Sports Car Club of America) Trans-Am sedan road racing. In fact the bore, stroke

Not all Fairlane fastbacks were backed by the factory, but many dealers backed racing cars. This 1968 Fairlane was raced in USAC in 1969 by Bill Behling of Milwaukee and sponsored by area dealer Northwestern Ford.

Benny Parsons took the ARCA title for Ford.

and displacement were the same as the Chevrolet Z-28's engine. In Fairlane form, no racing was intended as it only had a two-barrel carb, 9:1 compression and rating of 210 hp at 4600 rpm. Hydraulic lifters and mild stuff down under further relegated the 302 to bread-and-butter duties.

Back at it again was the 390, again in two versions. The two-barrel job lost five horses and now was advertised at 265 hp, while the four-barrel setup gained that many and was now rated at 325. At the start of the model year, the biggest engine was the Cobra 427 V-8. An ad for the mill said, "This is the old, rough, tough 427 forced into a shirt and tie."

The reference was to the hydraulic lifters (its first), milder cam and single four-barrel carb, which reduced the rating to 390 hp at 5600 rpm. To further the dethorning, the only way the 427 came was bolted to the C-6 Cruise-O-Matic. It was listed as an option for only the hardtops.

The 427's disgraceful condition didn't last long, as it was purged from the option list shortly after the start of the model run, never to return. Shortly after, the 289 returned to the lineup, rated at 195 hp and made the standard GT powerplant.

For a while, that left the 390 as the top engine, but not for long. Replacing the 427 as the performance powerplant in early spring was the Cobra Jet 428. The engine was a compromise of several factors including emissions, performance and adaptability to the average buyer. It wasn't available for wagons.

It used the 4.13-inch bore from the old 406, the 3.98-inch stroke from a Mercury engine, the 410, and a combination of parts to get decent performance. Heads came from the 406, but hydraulic lifters were similar to those on the 390, but would take higher rpm without pumping up. A 785-cfm four-barrel fed a cast-iron intake manifold. Valves were larger than the regular 428 that went into big Fords, but smaller than the 427. Compression worked out to 10.7:1. The bottom end was not the same as the 427, as two-bolt mains were used.

Perhaps the most interesting aspect of the 428 was the advertised horsepower. For years, Ford had been accused of overrating its engines, especially the 390's and early 428's. Ford kind of admitted that when Cobra Jet's rating of 335 hp at 5400 rpm was announced. The milder big Ford 428 was advertised at 340 at 4600, down five horses from the year before. Torque for the CJ also was weaker on paper at 440 pounds-feet at 3400, compared to 462 at 2800.

It wasn't that Ford suddenly had a streak of honesty, it was just that it would help CJ cars in drag racing classifications, as advertised horsepower was figured against weight.

While Ford had offered performance parts and packages for its big engines for years, it publicized and advertised the fact heavily in 1968 with a line of Shelby performance parts for the big-block FE engines. In the past, Cobra Kits were heavily promoted for smaller engines, but not much was done for the big ones.

New powerplants joined the 1969 Torino lineup. Ford Division General Manager John Naughton leans on a 428 Cobra Jet with the shaker scoop and appropriate identification. On the left is the 250-cube six, while the 351 Windsor V-8 is on the right. The Cobra Jet was installed in a few 1968 models; the 250 and 351 were new for 1969.

Since the introduction of the 1966 models, Fairlane had a shadow in the intermediate field, Mercury Comet, which shared chassis, basic body shell and engines. For 1968 that shadow grew larger and did as well, if not better, with its crack at the all-new styling. To make matters worse, Mercury would be horning in on Ford's racing efforts as well.

Mercury couldn't make up its mind what to call its intermediates for 1968. Its series used the Comet, Montego and Cyclone nameplates. When it filed its specifications with the Automobile Manufacturers Association, it used all three names instead of one of them. Since three of the five series had the Montego name attached and it eventually ended up using that name, it will be used here for 1968.

Body styles were the same as Fairlane, but distribution was different. At the bottom was the Comet formal roof hardtop, followed by the Montego series with four-door sedan and formal hardtop. Montego MX was the next step up with a four-door sedan, formal hardtop, convertible and four-door wagon. Another four-door sedan and formal hardtop comprised the Montego MX Brougham series, aimed at luxury. For performance people, the Cyclone and Cyclone GT were the ticket. Each had a formal and fastback two-door hardtop.

Montegos were wider and longer than the Fairlane, but shared the same wheelbases. Engines paralleled Fairlanes, with the 302 standard in all Cyclones and the MX Broughams at the start of the model year.

Without going into all the styling differences, the author feels that the Torino fastback was good looking, but the Cyclone was better. Perhaps it was because of its stock car racing history in the 1968 NASCAR season.

For the 1968 season, aerodynamics played a far more important role than things mechanical. Ford engineers were working on a new engine for racing, the 429. Development on the 427 had all but stopped, with the tunnel port design to be its most advanced state. Indeed, 427's were no longer being bolted into new cars by the time the racing season started.

There wasn't a whole lot to worry about anyway, for Chrysler wasn't doing much development on the Hemi either, telling NASCAR it was obsolete and needed the second carburetor Ford got.

Interest was concentrated on how the new bodywork for the Fords and MoPars would do. After the 1967 NASCAR and USAC seasons, it was hard to find any front-line team racing 1967 Fairlanes. While the outsides were radically different, the chassis and cage of the 1967 Fairlane and 1968 fastback were almost identical. About all the builders had to do was tear off the 1967 sheetmetal and put on the new stuff.

Cale Yarborough was the class of the 1968 Daytona 500 as he won the event in The Wood Brothers 1968 Mercury Cyclone. The Wood Brothers had fielded Fords for years, but were told to run a Mercury for the event.

Ford divided its teams between Ford Torinos and Mercury Cyclones in 1968. Going with Mercury was the Junior Johnson team with its driver LeeRoy Yarbrough. The front grille design of the Cyclone was more aerodynamic than the Torino. LeeRoy finished second in the 1968 Daytona 500.

As usual, the first appearance of the new cars came at Riverside in the Motor Trend 500. Ford was well represented with a fleet of Torino fastbacks, among them Dan Gurney who returned to Ford and The Wood Brothers.

Gurney seemed back at home in the new Torino as he put it on the pole with a record speed of 110.971 mph, then went on to keep Ford's and The Wood Brothers's streaks intact by winning the event in record time, with an average speed of 100.598 mph.

It was a Ford parade as Pearson was second in a Holman & Moody Torino, Jones took third in a Bill Stroppe Torino, Bobby Allison nailed down fourth in a Bondy Long Torino and Cale Yarborough made it a sweep of the top five positions in another Wood Brothers Torino.

Chrysler withheld its 1968 models for Daytona, as its drivers ran the 1967's at Riverside. Al Unser was the first across in a Rudy Hoerr 1967 Dodge Charger.

It wasn't until February that the aerodynamic superiority of the fastback shape would be felt. Cars set up for road courses, like Riverside, are fairly level to the ground. However, for superspeedway warfare, the rear ends are jacked up and the noses brought down. This setup slices through the air better when speeds get closer to the 200-mph mark.

When the Ford fastbacks were set up in this manner, the slanting rear surface became almost parallel to the ground, making an extremely aerodynamic device.

Tests at Daytona revealed a 1.5- to 2-mph advantage for the Mercury Cyclone fastback over the Torino. The grille on the Merc cut the wind better than the slightly recessed grille on the Torino. Wanting to be ready for the new MoPars, Ford brass decided to have both Ford and Mercury entrants in the Daytona 500.

On February 5, it was announced that factory Ford drivers would be David Pearson, A. J. Foyt, Bobby Allison and his younger brother Donnie. Wheeling factory Mercurys would be Cale Yarborough, LeeRoy Yarbrough and Mario Andretti. There also were some strong independent Mercurys driven by Tiny Lund and Jim Hurtubise.

In the past, Mercury had set up its own NASCAR teams and Ford had its teams. Now the two were being interchanged. Notable was The Wood Brothers being assigned to Mercury after being one of the pioneers in Ford's resurgence in stock car racing in the 1960's. Their driver was Cale Yarborough for the 1968 500. The association of The Wood Brothers with Mercury would be a long one. For all but a few races, they would field Mercury race cars right through the 1980 racing season. LeeRoy Yarbrough drove for Junior Johnson, who was with Ford off and on over the years. During the 1968 season he would use Fords on the short tracks and Mercurys on the longer ones. Holman & Moody split with Pearson in a Ford and Andretti in a Mercury.

The fastbacks' shape proved an advantage right from the start at Daytona, as they were running in the 190-mph range, about 8-mph faster than the fastest Fords at the 1967 Daytona 500.

When it came to official timing, Cale was the fastest with a record average of 189.222 mph in his Cyclone. A surprise was the second-fastest qualifier, Richard Petty in a 1968 Plymouth Road Runner. He had the only MoPar that could keep up with the Ford products and clocked 189.055. Driving the next-fastest Chrysler product was Al Unser at 183.525.

Petty's car attracted much attention, as his crew did a lot of aerodynamic work on it, including a special underside paneling. The flat-black roof also was different as the crinkled surface theoretically should have made the car slower, not faster than the traditional highly polished surface.

In the 500, the Mercury edge paid off with Cale Yarborough winning and LeeRoy Yarbrough taking second. Bobby Allison took third in a Bondy Long Torino, while Unser broke up the Ford Motor Company string with a fourth-place finish. Pearson put the Holman & Moody Torino into fifth, giving Ford four of the top five places.

There weren't as many factory-backed Fords in USAC stock competition in 1968. Shown are 1968 Torino fastbacks from the two main teams, Bill Stroppe's #15 being driven by Parnelli Jones and Jack Bowsher's #21, which he drove.

The finish was enough to make NASCAR give in and let MoPars run two four-barrels. Despite the help for the MoPars, the season was a good one for fastback drivers. Fords won twenty events, more than any other make with sixteen of those flags taken by Pearson, who won the national driving championship.

Mercury scored seven wins, six by Cale Yarborough, who also won the Atlanta 500, Firecracker 500 at Daytona and Southern 500 at Darlington, to help his winnings to $136,786, a single-season record. Petty scored sixteen wins to account for all of Plymouth's total. Dodge drivers took five flags.

The 1968 stock car season was also good for Ford elsewhere. A. J. Foyt drove Jack Bowsher's Torino to the USAC stock championship, scoring four of Ford's seven wins. Bowsher took the other three. Dodge drivers were a bit better, taking eight flags, while Plymouths took five.

Benny Parsons took the ARCA new car driving title, giving Ford Torinos three stock car championships in the 1968 season, a fact Ford advertised and publicized.

In drag racing, the 1968 Fairlanes were not the successes they were in stock car racing. It was not totally the fault of the cars, it's just that competition in the top super stock classes was increasing among the smaller pony cars. Now Ford Mustangs, Chevrolet Camaros, Plymouth Barracudas and Mercury Cougars were run by the best teams and had the best drivers.

Fairlane's mild 427 for 1968, plus increased bulk, made the 1966-67 Fairlane 427 still the faster way to go. The Cobra Jet 428 was used in drag racing, but mostly in Mustangs.

Perhaps Fairlane's greatest performance for 1968 was in sales. Production set a record of 372,327 units, a fantastic fifty-six percent above the 1967 run, especially considering that a strike stopped production for a few weeks. Industry production was also up, but less than ten percent. Indicating the role of the fastback in the popularity climb, the most popular model was the Torino GT fastback with 74,135 being built.

Montego production rose fifty-two percent to 123,113, but a total of 12,270 Cyclone fastbacks were made, or about ten percent of the total. This compares with a total of 106,587 Fairlane fastbacks of all types, or twenty-nine percent of the total.

1969

Fairlanes were now in a two-year styling cycle and the 1969 models produced only detail changes in the successful 1968 lines and trim. However, there were a considerable number of changes in the performance sector of the Fairlane lineup.

From the front a 1969 Fairlane Torino GT SportsRoof looked rather ordinary. It was the tail end styling that made it unique. This copy shows off the standard GT grille, side stripes, styled steel wheels and hood scoop with integral turn signals. The emblems on the scoop indicate a Cobra Jet 428 is beneath it.

Bodies were a rerun with minor trim changes. There were two new grille designs, depending on series. Chrome and stripes were rearranged, sedans and formal hardtops got new, smaller taillights that were more square than rectangular. Rear trim also was changed, varying per model.

Changes were greater under (and over) the hood. An increased stroke and partial redesign brought the six up to 250 cubic inches and a rating of 155 hp. The 289 disappeared again, this time for good from the Fairlane line, and the 302, rated at 220 hp was the next step up. Only one version of the 390 was around, the four-barrel 320-hp unit.

Back again was a refined version of the Cobra Jet 428. It was still rated at 335 hp, but the rpm level was down from 5400 to 5200. New cast-iron exhaust headers, a Holley 735-cfm four-barrel and larger intake and exhaust ports supposedly improved power, but didn't get Ford to budge its advertised output. Neither did the Ram-Air option for the 428, which involved an air scoop over the air cleaner that opened under full throttle so fresh, cool air could be pulled in. All horsepower and torque figures remained the same.

Ever since the 390 was added to the Fairlane line for 1966, there was a gap between the small-block and FE engine. Finally in the 1969 lineup, that was filled with the 351 Windsor.

It had the four-inch bore of the 289 and 302 and a new 3.5-inch stroke, but was not exactly the same block. It had 1.275-inch higher decks and the cylinders were spaced farther apart. Webs were stronger and took different intake manifolds and crank. Even the firing order was different.

The Windsor name came from the city where the 351 was manufactured, Windsor, Ontario, Canada. Actually the Windsor name was not in common usage in the 1969 models, because that was the only 351 available; but when a second version of the 351 came along for the 1970 models, it was used.

Speaking of names, even the 351 is an interesting choice. Bore and stroke for it were the same as Ford's FE big-block 352, which was available in big Fords from 1958 through 1966 models, four and 3.5 inches respectively. Actually it worked

For show car fans in 1969, there was the Super Cobra, based on the 1969 Cobra SportsRoof. New frontal styling, slats and a smoothed-out side were among the features. Note the nonstock wheels. The car was so popular it was shown again during the 1970 season.

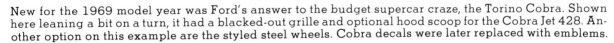

New for the 1969 model year was Ford's answer to the budget supercar craze, the Torino Cobra. Shown here leaning a bit on a turn, it had a blacked-out grille and optional hood scoop for the Cobra Jet 428. Another option on this example are the styled steel wheels. Cobra decals were later replaced with emblems.

out to 351.86 cubic inches, so rather than confuse customers and service/parts operations, it called the new engine the 351. Two versions of the 351 Windsor were offered to Fairlane buyers in 1969 models, a 250-hp two-barrel and 290-hp four-barrel unit.

Four-speed manual transmissions could be bolted to 351, 390 and 428 engines, while Cruise-O-Matics were optional for all engines with the unit used depending on engine size. Three-speed all-synchro manual boxes were standard through the 390.

Stealing the spotlight was the newest of the 1969 models, the Cobra. Available in formal roof hardtop or fastback (called SportsRoof for 1969), it was Ford's answer to the Plymouth Road Runner and its imitators.

Standard was no puny 383 like Plymouth, but the Cobra Jet 428 and four-speed transmission. That wasn't all. A competition suspension with staggered rear shocks (the left one ahead of the rear axle, the right, behind), six-inch wheels, F70x14 belted wide-oval white-stripe tires and racing hood pins were part of the package.

Outside was kept simple with wheel lip moldings, and Cobra insignia on the front fenders behind the wheel well and right side of the trunk or trim panel. The Fairlane 500 grille was done in flat-black. Inside were bench seats, but buckets were optional. Ford separated the buckets and console for 1969, making them individual options.

Cobras were the most expensive Fairlane hardtops for 1969, but considering the equipment included, the price was very reasonable. The SportsRoof started out with a list price of $3,164 and the regular two-door hardtop was $3,189. Prices were boosted later in the model year, but by less than $20.

To put the Cobra in perspective, *Motor Trend* tested a half-dozen supercars and put the Cobra up against some stiff competition, none of them the low-buck versions like the Cobra; and all of them carrying higher horsepower ratings.

First, the Cobra, which had Ram-Air, buckets and locking 3.5:1 differential, was listed with the lowest price, $3,945. *Motor Trend* also rated it best in quality and quietness.

The smooth lines of the 1969 Torino SportsRoof cut through the air, especially with the front end dropped and tail raised for stock car racing. Jack Bowsher's USAC stocker demonstrates that.

With a standard Cobra Jet 428 and four-speed transmission, the Torino Cobra SportsRoof version was Ford's answer to the Plymouth Road Runner budget supercar. This example has optional styled steel wheels with plain centers. The decals on the front fender and rear panel were later replaced with emblems.

In performance, it ranked third behind a pair of 440-cube MoPars, a Plymouth GTX and Dodge Charger R/T. The Cobra did 0-60 in 6.3 seconds, compared to 5.8 and 6.1 seconds respectively for the GTX and Charger R/T. In the quarter the snake hit 14.5 seconds and an even 100 mph, compared to 13.7/102.8 for the Plymouth and 13.9/101.4 for the Dodge.

However, the Cobra blew away a Pontiac GTO, Chevelle SS-396 and Buick GS 400. Sure, the GM cars had smaller engines, but it encouraged the street runners to think twice about the 1969 models.

Unfortunately the Cobra was a solid first in one category, poor gas mileage. Although many suspected the car wasn't properly tuned, it still got only 7.7 to 9.1 mpg.

Elsewhere in the Fairlane line, models were the same as the year before. Torino GT models resumed having the 302 standard. A simulated hood scoop with directional indicators in the back was now standard, but production problems caused some early models to be built without them. The SportsRoof GT came with a revised C-stripe on the side which sent the upper stripe along the middle of the side panels. Torino and Torino GT models got a different grille than the other models with a center bar and plastic construction. The others got a unit with twin extruded aluminum bars and smaller bars in the background.

Fairlane faced its usual onslaught of hot competition in the intermediate range with all models also getting facelifts to one extent or another. However, one particular manufacturer started a war of aerodynamics that was to get out of hand.

In the 1968 stock car battle on the superspeedways, Fords and Mercurys were the undisputed rulers, but it wasn't supposed to be that way. The Dodge Charger

A mildly restyled grille, rearranged side trim, new wheel covers and change in the rear side-marker light indicate this Fairlane 500 convertible is a 1969 model.

Bobby Unser won the USAC-sanctioned Pike's Peak Hill Climb July 4, 1969, in a 1969 Torino SportsRoof owned by Bill Stroppe and backed by Ford. He set a record for the trip up the mountain.

looked like it should be just as fast, but it wasn't. A closer check showed the MoPar's recessed grille and notchback rear window with sail panels didn't cut a smooth path through the air at high speeds.

Dodge countered with the 1969 Charger 500, a specially modified hardtop with a flush grille, fixed headlights and slanted rear window that was molded to the outer edge of the sail panels. This would put Charger on par with the fastbacks, Chrysler figured.

The only problem was, Dodge couldn't wait to tell the world what it did and made the announcement in fall of 1968. Ford knew what its target was and had its answer to the 500 announced in time to be legal at Daytona, the Torino Talladega SportsRoof.

Talladega was the name of a small town in Alabama where the largest superspeedway in the country, 2.66 miles, was scheduled to open during the 1969 season.

The Talladega sales brochure started off, "Ford powered cars have dominated the International performance and racing field for the past five years and 1969 is not to be an exception."

It was right, and here's what was done to assure that leadership in NASCAR: While the fastback part of the 1968 Torinos was fine, it was no secret the frontal area could be cleaned up. The accidentally faster Mercury Cyclones were an example. Ford wanted to win races, but the cost of restyling the front of all Fairlanes to do it was not worth it. However, rules said that if 500 of any model used for racing were made, it would be legal. Hmmmm.

Though there is arguing over who did the designing, Ford or racers, the most notable feature of the Talladega was the extended front fenders and tapered panel in front of the hood to a flush Cobra grille. For a bumper, the rear unit was used. It was cut in the center, trimmed and welded together in a slight vee, serving as an air dam for the front end.

Directional signals and side lights were wiped out with the extended fenders, so truck units were used behind the grille for the directionals and the side light was taken care of by a yellow unit like the rear red one. Another feature with superspeedways in mind was rocker panels, which were rolled under a bit, reducing total body height to 52.6 inches from a stock 53.5. Length of the Talladega was 206 inches, up 5.9 from the regular models.

Ford was not out to make the Talladega a popular model that could be custom ordered by dealers. It had to have the 500 units made well before the February 20 Daytona 500 to be legal, and since Talladegas took special work to build, it wanted to get the project over with.

When A. J. Foyt's Torino blew an engine in a 1969 USAC stock car race practice session, the Jack Bowsher crew went to work yanking out one Boss 429 and installing another. With factory backing, the team was well-equipped for the job.

Standing room only greeted the Ford Drag Team in a 1969 performance at a Ford dealer. Here factory driver Hubert Platt explains the Ford performance parts available. The teams went to dealers between weekend drag races. Platt was part of the Eastern team.

So Talladegas were rolled out of the Atlanta plant in a short period of time in late January and early February, all with pretty much the same equipment. Counting prototypes (not made in Atlanta) there were 754 Talladegas made.

Despite prototypes having an air scoop and Ram-Air listed as an option, production Talladegas came with the regular Cobra Jet engine and C-6 Cruise-O-Matic transmissions. They also came with power steering and disc front brakes, competition suspension, F70x14 white-stripe tires on six-inch styled steel wheels. Hoods were flat-black and colors for the production cars were Wimbleton white, presidential blue and royal maroon. Inside was nothing to get excited about, as black interiors had cloth-and-vinyl bench seats.

Production figures listed for 1969 models will not list the Talladega, or the Cobra for that matter, as both carried the same body number as the Fairlane 500 SportsRoof; and in the case of the Cobra, the formal roof hardtop. Despite this they were priced and marketed as separate models.

NASCAR was satisfied that enough Talladegas were built and the car was homologated (made legal) for Daytona, but that was only half of the Ford NASCAR story for 1969.

To power the Talladegas, Ford wanted to legalize a hemi-headed version of the new 385 series big-block 429 it had been installing in Thunderbirds since the 1968 models and big Fords starting with the 1969's.

For some reason it was decided to install the engine in 500 special Mustangs, which were built at Ford's Kar Kraft specialty house and called Boss 429's. At the time NASCAR no longer required that engines be installed in the same car in which they were raced.

Most popular of the 1969 Montego hardtops were the formal roof models which outsold the fastbacks by a large margin.

Ford had the Torino Cobra and Mercury the Cyclone CJ 428 in the budget supercar sweepstakes for 1969. Standard were the CJ 428, four-speed, hood pins and special decals. This example has optional Ram-Air induction and turbine-styled wheel covers.

The Boss 429 run wasn't approved by NASCAR in time for the Daytona race, so the Talladegas had to use the old 427 wedge, which wasn't a production engine for the Fairlane at all, since it was last built for new cars late in 1967.

If you can figure all that out, then you're ready for a career in politics.

Before we get into the NASCAR fun and games, we can't forget Mercury, which was shadowing Ford's every move.

First, it simplified its Cyclone intermediate series, kicking out the formal roof hardtops. Also gone was the Cyclone GT. The hot version was the Cyclone CJ which, like the Cobra, came with the 428 standard. Buckets were optional in both Cyclone and CJ versions. Incidentally, the Cobra Jet 428 was called CJ 428 under the "sign of the cat."

With the Cobra covered, Mercury's next move was to take care of the Talladega. At the end of December 1968, Mercury announced the Cyclone Spoiler. Here is where a bit of confusion sets in. The press release talked only of a Spoiler model and came with photos showing a Cyclone Spoiler with regular front fenders and a retouched photo with a Talladega-like front end—only with a one-piece hood extending down to the grille.

The release, however, mentioned only one Spoiler and called attention to its special features, the most obvious of which was a twin strut-mounted spoiler attached to the trunk.

As things were sorted out, there would be two Cyclone Spoilers for 1969, a regular-length job and one with a Talladega-like nose, which would be called Spoiler II. The cars came in two editions: the Cale Yarborough Special with a metallic-red roof similar to The Wood Brothers colors and a Dan Gurney Special with a metallic-blue roof. Gurney was under contract to Lincoln-Mercury that year; but it was interesting that his name be on a Mercury, as all his big stock car wins were in Fords.

The bottoms of the cars were white with appropriate twin stripes on each side and decals for the respective drivers. A total of 519 Spoiler II's were made, all with 351-4V engines.

The 1969 season turned out to be a high-water mark of identification between racing in general and NASCAR in particular and the Ford sales-promotion forces. Ford dealers were encouraged to set up "performance corners" in their showrooms and feature the performance cars.

Noses were different on the 1969 Torino Talladega (above) and regular 1969 Torino. The extended Talladega nose and flush grille were more aerodynamic than the recessed standard unit.

Torino Talladegas appeared in the 1969 Daytona 500, but did not carry Talladega lettering due to a dispute over use of the Boss 429. They had 427 power. Factory-backed examples include the #11 Jack Bowsher Talladega driven by A. J. Foyt and the #17 Holman & Moody Talladega driven by David Pearson.

The May 1969 *Ford Dealer* magazine pointed out the value of performance, "Performance is the biggest, most talked about selling feature among today's youth-oriented buyers who form a nucleus of the muscle market representing approximately 600,000 cars. That's a volume industry bigger than the entire luxury car market. Moreover, consider the growing influence that these 600,000 buyers have on the total market."

To make sure it had the NASCAR situation well in hand, Ford signed two more big guns for its arsenal in 1969. The biggest was Richard Petty, the winningest NASCAR driver, who had driven Plymouths since the middle of the 1959 season.

Petty felt a Charger 500 would be more competitive with the Ford fastbacks than his notchback Plymouth Road Runner. Chrysler said no, he had to run Plymouths, so Petty said no and signed with Ford.

In the course of the season, Ford also signed Smokey Yunick, who had given Ford fits over the years with his fast Pontiacs and Chevrolets. Yunick had a reputation for building fast and technically advanced cars, then hiring top drivers, and had signed originally with Ford in 1957, just before the factory pulled out. He was no doubt brought back by Ford President Bunkie Knudsen, who came over from GM.

It's interesting to note that Yunick was fielding Pontiacs when Knudsen was at Pontiac, and Chevrolets when Knudsen headed that division.

At the time, NASCAR started its season at the end of the previous calendar year with short-track events. Two late 1968 calendar events that counted to the 1969 points went to Plymouth drivers. Petty won a 250-miler November 17 and Bobby Allison took the December 8 hundred-mile contest at Montgomery, Alabama. They would be the only 1969 season wins for Plymouth.

First event in 1969, as usual, was the rain-delayed February 1 Motor Trend 500 at Riverside. Petty started his association with Dearborn in calendar 1969 with a bang and won the race. His accomplishment was somewhat amazing for despite all his talents, Petty was known as a terrible road racer. Lots of practice changed things.

Ford drivers also took second and third with A. J. Foyt runner-up and David Pearson's car taking third with driving relief from Parnelli Jones. Neither the 429

When NASCAR officials visited the Atlanta Ford plant to check if enough Talledagas were made to qualify for racing, Ford folks put on a show for them, using 31 cars to form the numbers 500 and the letter T. Note all had hoods void of the air scoop of the prototype.

An extended front end made the mid-year Torino Talladega the longest Fairlane for 1969. Built so NASCAR stock car racers could use it, the Torino Talladega had several unique features including a modified rear bumper up front, hidden turn signal lights, rolled rocker panels and a flat-black hood. This preproduction model has an air scoop which the production cars didn't have.

nor Talladega was legal to run, so Fords carried 1969 trim and the lettering "Torino Cobra" on the side.

The February 23 Daytona 500 would prove to be a combination of frustration and triumph for the Ford armada.

Talladegas were legalized, but not the 429, so the 427 had to be used to fight the aerodynamically improved Dodges. Ford ordered that its factory cars retain the Torino Cobra lettering on the Talladegas, insisting that a true Talladega had a 429 (which it didn't). Also, the Cyclone Spoiler II was not homologated in time for Daytona, so all factory teams got Talladegas.

Pearson put his Holman & Moody Talladega on the pole and won the first 125-mile qualifying race. The Dodges were very close to the fastbacks this time, as all cars now had to have a single four-barrel carb. Bobby Isaac took the second qualifying race in a 1969 Charger 500.

The race itself was a see-saw battle between the Fords and Dodges with LeeRoy Yarbrough taking the lead from Charlie Glotzbach's Dodge on the final lap for a Ford win. Donnie Allison and Foyt brought Fords in third and fourth, respectively, and Buddy Baker was fifth in a Dodge.

Equality was not the way Ford liked to deal with things. Both the Boss 429 (it had several other names) and the Spoiler II were legal for the March 30 Atlanta 500 and Cale Yarborough took advantage of both, driving a Wood Brothers Spoiler II 429 to victory. Pearson was right behind, leaving skeptics who said the engine wouldn't work out of the box, silent.

Not silent was Dodge. Its Charger 500 wasn't getting the job done, but that didn't mean it had to stop there. On April 13 Dodge shocked the racing and automotive worlds with its answer to the Talladega and Spoiler II: the Charger Daytona.

Using the basic Charger 500 body, stylists grafted on a long pointed nose and hung a high stabilizer wing on the back. It would be awhile before the cars were legal, but it was clear that an aerodynamic war was on.

Meanwhile the 429 was working well and LeeRoy Yarbrough was the hottest superspeedway driver on the circuit. Driving Junior Johnson's Mercury he won the May 10 Rebel 400 at Darlington, May 25 World 600 at Charlotte and July Fourth Firecracker 400 at Daytona; only in the latter event, was he back in a Ford.

Ford was worried that Dodge would catch it in the battle for the manufacturer's championship and switched the red-hot Johnson team over to the Ford column, just in case.

The change made little difference, as Yarbrough continued his winning ways with the August 10 Dixie 500 at Atlanta, September 1 Southern 500 at Darlington and October 26 American 500 at Rockingham for a record seven superspeedway races in a single season. His winnings of $188,605 was also a single-season mark.

Ford won the manufacturer's title, Pearson repeated as the driving champion and in all Fords took twenty-six wins, four more than Dodge.

Mercury's answer to the Ford Talladega was the Cyclone Spoiler II. This photo was issued by the factory and featured a retouched front end, which was not correct, as the hood extended down to the grille. Actual models had a nosepiece like the Talladega. The decal on the front fender indicates a Cale Yarborough Special, meaning metallic-red paint on the roof.

A mid-year addition to the 1969 Mercury Cyclone lineup was the Spoiler. Shown wearing out the rear tires is a Dan Gurney Special, which meant a blue roof. The front end on this example was standard Cyclone, not extended like the Spoiler II. The device on the rear deck was responsible for the name.

Dodge's Charger Daytona made its debut in the September 14 Talladega 500. A dispute with the drivers over track conditions resulted in most of the big name drivers not competing and all the factory Fords and Mercurys being parked. Richard Brickhouse drove a Ray Nichels Daytona to victory.

So, the Talladega first appeared in the Daytona 500, and won; and the Dodge Daytona made its debut in the Talladega 500, and won.

In USAC stock competition, where aerodynamics have a less important role, Dodge and Ford drivers battled it out for the most wins with Dodge winning, ten to nine. The Charger Daytona was not legal in USAC in 1969.

The battle for the driving championship, however, went to Roger McCluskey, who drove a Norm Nelson Plymouth to five wins and edged Foyt, who drove a Jack Bowsher Torino, in points.

One of Ford's wins came at Pike's Peak when Bobby Unser wheeled a Bill Stroppe Torino to victory. On July 14, Unser also won the Miller 200 at Milwaukee. His car carried Torino Talladega lettering, but the front end was of conventional Torino configuration, so in 1969, Talladegas were lettered as being Cobras and Cobras were lettered as being Talladegas.

NASCAR and USAC were only part of the "Going Thing" performance program for 1969. Ford expanded its Drag Club program, forming local clubs all over the U.S. and Canada. Members got a twenty-five-percent discount on parts.

To back up the Drag Club activities, Ford had two drag racing teams touring the United States, running at various drag strips. Between race weekends, they would visit local dealers and hold performance seminars. They campaigned a Cobra and Mustang Mach I, each powered by the Cobra Jet 428. Covering the East were Hubert Platt and Randy Payne, while Ed Terry and Dick Wood took care of the West.

As is usually the case in the second year for a style, sales of the Torino dropped for 1969, but not all that much, as production eased a little over one percent to 366,911. Unfortunately, a strike hurt the 1968 total, which was a record, while 1969 cars were produced all model year. The Torino GT SportsRoof was again the most popular.

1970

Ford was deeply into its longer-lower-wider kick when the time came to design the 1970 intermediates. They grew in every dimension except height.

Also, there was some controversy as to what Ford intermediates for the 1970 model year were called. Literature referred to them as Torinos, but documents

After the Boss 429 motor was legalized in NASCAR part way into the 1969 Grand National season, factory Talladegas carried the Torino Talladega name on the rear quarter panels. Hoods also carried the new engine size. Shown at speed are David Pearson (#17-Holman & Moody Talladega), Donnie Allison (#27-Banjo Matthews Talladega) and Richard Petty (#43-Petty Talladega).

filed with the Automobile Manufacturers Association called them Fairlanes. To further confuse the situation, a series was added mid-year to replace the Falcon called 1970½ Falcon. It was listed with AMA as being a Fairlane, marketed as a Falcon and late in the model year redesignated a Torino.

Since the Torino name pops up more than any other, we'll go with that one.

Underneath, the Torino for 1970 was pretty much like the 1968-69 models which had a similar relationship to the 1966-67's. Wheelbase grew an inch to a new high, 117. Wagons and Rancheros also added an inch to hit 114. Treads were widened so wheels wouldn't look lost under the wider bodies. The front went from 58.8 to 60.5 inches and the rear grew from 58.9 to an even sixty inches.

Plunked on top was a completely restyled body, except for wagons which got the beltline down treatment again. Overall length grew from 201.1 to 206.2 inches for non-wagons. Width picked up two inches or so, depending on model and was in the neighborhood of 76.5 inches, again depending on the model.

Height also varied per model, but as an example, the SportsRoof dropped from 53.5 to 51 inches. Weights gained more than one hundred pounds, continuing the expansion in that category.

Styling lost the simpleness of the 1968-69 models and was tabbed by the ad copy writers as being "Shaped by the Wind." That phrase was to backfire come the racing season.

Sleekest looking of the 1970 Torinos was easily the SportsRoof two-door hardtop. Lines started with the new 1970 windshield which was swept back six degrees farther than the earlier models. The rear quarter window slanted toward the back of the car, but was made smaller by the rear quarter panel, which blended into the roof. Lines then swept back from the rear edge of the roof to the end of the trunk where they rose for a spoiler effect.

It should be noted that the rear line from the roof edge was not flat like the previous models, but rather curved down, then back. The rear window also took this curve. An option for the rear window was Sports Slats in flat-black plastic which were supposed to add a sporty flair and protect rear seat passengers from sun at the same time.

The conventional two-door hardtops were similar to the 1968-69 formal-roofed hardtops in concept; more sweeping lines with the roof going back farther into the trunk. When the Falcon two-door sedan was added mid-year, it used the hardtop style with rear window and door frames, and a center post.

Grille styling varied with each model, but top-line cars got optional disappearing headlights. Grilles were recessed and pointed at the center. The front fenders and bumper also met in a point.

Hoods and the cowl were redesigned so the wipers could be hidden. A fresh-air inlet was at the rear of the hood, which now extended from the grille to the windshield.

A multitude of lines marked the sides, all allegedly put there by the wind. Rear quarters bulged a bit, as was fashionable, but were of a different line altogether

"Shaped by the wind" the ads said about the 1970 Torino GT SportsRoof. Visually one would have a hard time arguing with that, but the wind knew better and proved the 1970 models to be considerably less aerodynamic than the 1969's.

on the SportsRoof. Taillights filled the ends of the panel between the bumper and the trunk lid, with differences in trim between models.

At either end of the new Torinos were valance panels, or splash pans, below the bumpers. In the past, Fairlanes were content with big bumpers to define the bottom of the body ends.

Also growing for 1970 was the model lineup. Starting the year out at the bottom was our old friend, the Fairlane 500, making its final appearance. It came in four-door sedan, two-door hardtop and four-door wagon.

One step up came the Torino, with a four-door sedan, two-door regular hardtop, new four-door hardtop (the first year for the body style in a Ford intermediate) and four-door wagon. Mid-year a SportsRoof two-door hardtop was added.

Aimed at the luxury buyer was a new series, the Torino Brougham. Two- and four-door hardtops and the Squire wagon populated the series and all came with a V-8 standard.

Now, sports fans, here come the performance cars. Torino GT came in SportsRoof and the Torino's only convertible model, with a 302 V-8 standard. Back for an encore was the Torino Cobra, in the new SportsRoof form, complete with a big engine standard. This time it even got its own model number, 63H. The addition of 1970½ Falcons brought another four-door sedan, wagon and the only two-door sedan offered that year in the intermediate body style.

A closer look at the performance-oriented models is in order, starting with the Cobra, which followed the same basic formula as the year before. It looked

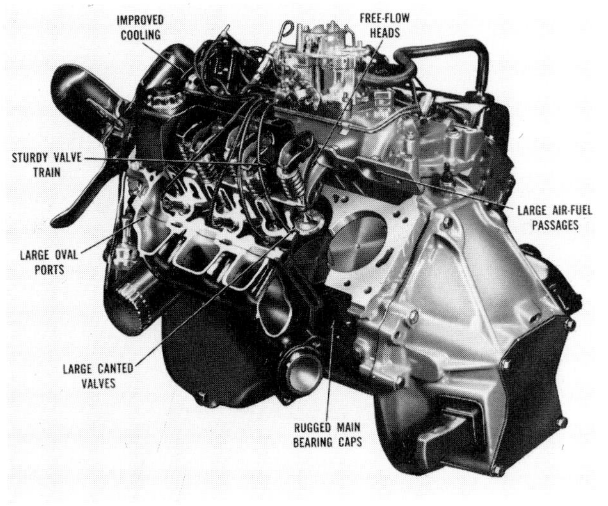

Though it had the same displacement, bore and stroke as the 351 Windsor, the 351 Cleveland was a different engine. The first member of Ford's new 335 series, it had free-breathing heads, staggered valves and a strong underside. It was first offered in 1970 models and initially came in both two- and four-barrel versions.

A mid-year addition to the Torino series for 1970 was the SportsRoof. Previously the style was only available in Torino GT and Torino Cobra form.

mean in its new suit of clothes. The SportsRoof hardtop came standard with a flat-black centered hood, complete with new-style hood pins, a blacked-out grille with fixed headlights and a chrome bar in the middle, wheel lip moldings, Cobra decals or emblems (which replaced the decals) on the front fenders and rear trim panel and a rather plain interior.

Mechanically it maintained its muscular stature with a new 360-hp 429 standard, along with the tough Ford four-speed, stirred by a Hurst shifter (but not linkage). Competition suspension came as part of the package, as did seven-inch-wide wheels and F70x14 raised-letter tires. Prices started at $3,249 and were raised in December to $3,270, leaving the Cobra in the very reasonable category.

Torino GT's turned out to be the show cars of the line. The fixed or optional disappearing headlights were set in a full-width egg-crate grille with a round GT medallion in the center. A nonfunctional scoop took up the middle of the hood. Side trim included wheel lip and rocker panel chrome, plus GT letters and five chrome slashes just behind the door that begged to be part of the Laser Stripe reflector tape that was optional and extended from the five strips forward.

GT taillights were set behind a honeycomb pattern similar to that on the Lamborghini Marzal show car and extended all the way across the back, thanks to a reflector panel in the middle.

Bench seats were standard for both the GT and Cobra, with high-back buckets and console optional. The dash was redone again, this time with a horizontal speedometer and optional ribbon tachometer.

Cale Yarborough was all smiles looking over a 1970 Mercury Cyclone, but he spent most of the 1970 season racing a 1969 Spoiler II.

Mercury didn't get a copy of the SportsRoof for its 1970 Montego line, but instead had three special versions of the regular two-door hardtop: Cyclone, Cyclone GT (shown) and Cyclone Spoiler.

Ford didn't restrict its changes to just the styling, as two new engine designs were under the shapely hood. Gone were the last of the FE series big-blocks, the 390 and 428. Back was the 250 six, which was standard on Falcons, Fairlane 500's and Torinos, and the 302 V-8, which was the standard engine for the GT's and Torino Broughams.

Even though the size was the same, Torino's 351-cube V-8 was new for 1970. In 1969 the Windsor V-8 came in two versions. For 1970 the new Cleveland block was available in two forms, 250-hp two-barrel and 300-hp four-barrel. However, it is likely that some Windsor two-barrels with the same rating found their way into Torinos.

The four-barrels were Clevelands, named for the plant where they were constructed. The Cleveland engine used the four-inch bore and 3.5-inch stroke of the 351-W but the block and heads were of different design. An easy way to tell is that the Cleveland block had an extension in front to house the timing gears and chain and also mount the fuel pump. The Windsor block used a cover for the system.

Heads on the Cleveland engines used canted valves like the Boss 302 and 385 big-block series. The lessons learned in racing were put into practice with extremely free-breathing head design.

Two- and four-barrel Cleveland engines for 1970 differed in more than intake manifold design. Four-barrel heads had bigger intakes and valves, and a hotter cam.

Four-barrel blocks were supposed to come with four-bolt main bearing caps, but some came through with two. The caps differed from those on the old cross-bolted 427's in that both bolts went straight down into the webs on the 351-C, while the 427's second set of bolts came from outside the block. Also the 351-C four-bolts were on all five mains, compared to the center three on the 427.

The number of versions of Torino's biggest engine for 1970, the 429, depends on how you look at it. First, it wasn't the same 429 that had gone into the 1969 NASCAR stocks. Those were Boss 429's with semi-hemi aluminum heads (the Boss 429 was listed for the Torino for 1970 at the very start of the model year, but few if any were built). Torino 429's had more conventional cast-iron heads with canted valves. They were part of Ford's fairly new 385 engine series, which had thin wall construction and no skirts.

Mildest form of the 429 was the 360-hp version standard on the Cobra. Its horsepower rating was 360 at 4600. Compression came to 10.5:1 and two-bolt mains were underneath.

With the 428 gone from the lineup, Ford transferred the Cobra and Cobra Jet titles to the hotter 429 versions.

Next step up from the 360-hp was the 370-hp Cobra edition of the 429. It came with a 700-cfm four-barrel, 11.3:1 compression and two-bolt underside. When fitted with the shaker hood scoop, which was attached to the air cleaner instead of the hood and gyrated with the motions of the engine, the engine was known as the Cobra Jet Ram-Air V-8. Advertised output didn't change, with horsepower remaining at 370 at 5400. Torque was also the same as the Cobra at 450 pounds-feet at 3400 rpm.

However, the hottest 429 short of the Boss was obtained by ordering the Drag Pack option. With that you got the Super Cobra Jet 429 with four-bolt mains on the center three webs, solid lifters, 780-cfm Holley four-barrel carb, cast-iron headers, solid lifters, free-breathing heads and a host of other items to bring the 429 up to 427 competition standards.

Factory Ford teams were faced with a dilemma during the 1970 season. To be competitive on the superspeedways, they had to run year-old 1969 Talladegas (top). The 1970 SportsRoof and hardtops could be used on the shorter tracks (bottom). Corporate labeling policy eliminated the Talladega reference on the 1969 quarter panels and used the Cobra designation on the 1970's. Shown are the cars used by Holman & Moody (#17) driven by David Pearson.

The Super Cobra Jet was underrated with an advertised output of 375 at 5600. It came with or without Ram-Air. Also included in the package were 3.91 or 4.30 locking differentials and an oil cooler for the engine. Transmission availability for all 429's was the Ford four-speed or SelectShift Cruise-O-Matic. The shaker hood scoop was also optional with the 351-4V (Cleveland) V-8, but also had no effect on the advertised output.

Back in 1963, the stock car and drag racing sanctioning groups put a seven-liter (about 430-cid with tolerances) limit on engine size to stop a parade of cubic inches that was rumored to be heading past the 500 mark. This held a lid on size for a while, but it didn't last long. Starting in its bigger 1966 models, Chrysler brought out a 440-cid version of its B-block wedge. At first it was only for the family cars, but that changed with the 1967 models, which saw 375-hp versions of the 440 in the MoPar intermediates.

The new size was readily accepted in drag racing and Chrysler had an advantage. NASCAR and USAC stuck with the former limit, but it didn't make any difference to Chrysler for it had its 426 Hemi for those battlegrounds.

During the 1969 model year, "Six Pack" options for the MoPar 440 were announced with 390-hp ratings. Six Packs were triple two-barrel carbs.

At the time, Ford was doing its performance development work on an engine that was still within (more or less) the seven-liter limit. General Motors was ready to end its ban on big engines for intermediates and starting with the 1970 models, would offer a bunch of 454/455-powered cars.

Before the model year was over, you could get an SS-454 Chevelle with rating up to 460 hp, Pontiac GTO with a 455 at a mild 360 hp (the 370-hp Ram-Air 400 was hotter), 370-hp Olds 4-4-2 455 or Buick GS Stage I with a 360-hp 455.

Except for the ultra-hot Chevelle LS engines, the Ford's lack of inches didn't keep it out of the running, at least on the street.

Motor Trend was muscle car comparing again and in one test put an SS-454 Chevelle, Cobra Jet Cobra and Road Runner with six barrels together. All had the trick scoops that abounded that year. The Chevy and Ford had automatics, the Plymouth a four-speed. In a 0-60 run, Chevy and Ford tied at six seconds; the Plymouth was slower. Ford had the top speed in the quarter, 100.2 mph, to 99 for the Plymouth and 97.5 for the Chevy; but lost out in elapsed time with 14.5 seconds to 13.8 for Chevy and 14.4 for Plymouth. The testers suspected problems with the Plymouth, had it retuned and reran the test, resulting in faster figures, but did not retune the other two cars.

All of Torino's competition was not in the performance end of the market. The luxury intermediate was also becoming a factor. Pontiac brought out its Grand Prix on a stretched intermediate chassis for 1969 and Chevrolet followed suit with its 1970 Monte Carlo, on a 116-inch wheelbase intermediate chassis.

Ford's answer was a plush Torino, the Brougham. Since Chevelle two-doors were on a 112-inch wheelbase, Torino was closer in size to the Monte Carlo anyway, so no need was seen for a separate model.

Mercury redid its Montego lineup for 1970, but for once didn't do everything Ford did. There was no Mercury fastback. Performance models were still called

Longer, lower and wider described the 1970 Torino SportsRoof. This GT shows the optional hidden headlights and Magnum 500 wheels. The nonfunctional scoop came standard.

Cyclones, but were based on the regular two-door hardtop. There were three versions, Cyclone, Cyclone Spoiler (spoiler on the rear deck) and Cyclone GT. Engines available were similar to Torino.

The war of aerodynamics was by no means over even though Dodge did not mass produce a 1970 version of the Daytona. It didn't need to, the 1969 Daytona would be legal for two more racing seasons. There was a new winged warrior from Chrysler, however, the (take a deep breath) Plymouth Belvedere Road Runner SuperBird. SuperBirds, as they were called for short, used the trick nose and wing from the Daytona and adapted them to a sleeked-up Road Runner two-door hardtop body.

Rules changed for homologation and now one had to be made for each two dealers, putting Plymouth's production goal past the 2,000 mark.

While the Daytonas were kept in line by the 1969 Talladegas and Spoiler II's, something new would be needed for 1970—and so came Ford's proposed successor to the Talladega, the King Cobra. If Chrysler could graft a superspeedway nose on a stock intermediate, why not graft a new clip from the firewall forward?

Using a design similar to the Ford GT sports racing cars, the King Cobra had a long hood sloping down to a minimal front bumper with sloping fenders to match. The new front end was grafted on a 1970 SportsRoof hardtop. The rear two-thirds of the car was more or less stock. It *looked* sleeker than the winged things from Chrysler.

The cover of the October 1969 issue of *Motor Trend* featured the King Cobra, as did a color section inside. It looked like Ford had the latest word in the aerodynamic supercar battle.

However, a number of events took place in a short period of time that not only put a stop to the King Cobra but before too long halted most of Ford's racing program.

Most significant was the dismissal of Bunkie Knudsen as Ford president. He was a proponent of performance and racing on many fronts as a way of promotion. When Knudsen left, so did Ford's racing drive. Henry Ford II may have been enthused about Total Performance in the middle 1960's, but with greatly increasing costs, especially for the onslaught of federal regulations, Henry's priorities were elsewhere than the winner's circle.

More bad news came from the racetrack. First, tests of the 1970 SportsRoof revealed that despite the advertising claims the new style was anything but shaped by the wind. In fact the 1969 Talladega was up to ten miles an hour faster than the 1970 model.

The King Cobra was no answer either, as the front end plowed and needed a rear wing if it was to do anything at all. With production requirements now higher and Knudsen gone, the King Cobra was let die. A few were made and two are now known to be in the hands of collectors, race car builder Bud Moore and Steve Honnell.

So, if Ford wanted to do anything at all in NASCAR, it would have to run the 1969 models, which looked far different from those being sold and advertised as being aerodynamically efficient.

In the middle of all this, an announcement was made that the racing budget was being cut in all areas for 1970. Advertising would also be toned down, bowing to the safety folks. "We're going to continue on all fronts, but on a reduced basis," said Lee Iacocca, head of Ford's North American operations.

One of the two 1970 King Cobras known to exist is owned by Steve Honnell. He bought it in pieces from Holman & Moody and has restored it to new condition. It is yellow with black hood paint and side stripes. It looks just the way the King Cobra appeared on the cover of ***Motor Trend***. Not exactly stock under the hood, it has a 494-ci Can-Am engine. The King Cobra was basically stock from the firewall back, with a fiberglass front clip. The other known King Cobra is owned by stock car builder Bud Moore.

Chrysler was ready to fill in part of the void. Using the SuperBird and money as bait, it got Richard Petty back in a Plymouth. It also picked up Dan Gurney for Plymouth, giving him a SuperBird for Riverside and Barracudas for his Trans-Am effort.

With aerodynamics not as much a factor, Ford teams arrived with 1970 models for the January 18 Motor Trend 500 at Riverside. The new cars would be used for other short-track races and a few teams tried them on superspeedways during the year.

USAC drivers were the stars at Riverside. Parnelli Jones put The Wood Brothers 1970 Cyclone on the pole, as regular driver Cale Yarborough was recovering from a broken shoulder from the final 1969 event at Texas.

Ford's win streak was extended to eight straight Motor Trend 500's when A. J. Foyt wheeled Jack Bowsher's 1970 Torino to victory. Right behind was fellow USAC star Roger McCluskey in Norm Nelson's new SuperBird. Then came LeeRoy Yarbrough and Donnie Allison in new Torinos and Petty in a SuperBird.

For Daytona it was back to the 1969's for Ford and Mercury teams, but no one was feeling sorry for them after Cale put The Wood Brothers Spoiler II on the pole at a record 194.015 mph, healing shoulder and all. Cale then came back to win the first 125-mile qualifying race and Charlie Glotzbach took the second in a 1969 Daytona.

The February 22 Daytona 500 went to newly acquired Petty teammate Pete Hamilton in a SuperBird. David Pearson was second in a Talladega, followed by three Daytonas, driven by Bobby Allison, Glotzbach and Bobby Isaac.

Overall the 1970 season belonged to Chrysler. Plymouths took twenty-one events and Dodges seventeen. This compared to six for Ford and four for Mercury.

However, the year was not without its high points for Ford fans. Six superspeedway races went to 1969 FoMoCo products. David Pearson took the May 9 Rebel 400 at Darlington; Donnie Allison, with help from LeeRoy Yarbrough in relief, claimed the May 24 World 600 at Charlotte; Cale Yarborough gave Mercury a win in the June 7 Motor State 400 at Michigan International Speedway; Bobby Allison scored again in the July Fourth Firecracker 400 at Daytona; LeeRoy took the September 11 National 500 at Charlotte in Junior Johnson's Mercury; and Cale closed out the superspeedway season with a November 15 flag in the American 500 at Rockingham.

As the 1970 season closed, it was announced that Ford was withdrawing from most forms of racing participation. The word came down on November 20 from Matthew S. McLaughlin, vice president of the sales group. Only limited drag and off-road support would continue. For NASCAR this meant that teams wanting to run Ford products were on their own.

Chrysler also cut back its programs with word that only a Petty Dodge and Plymouth would have limited factory support for the 1971 NASCAR season.

NASCAR, sensing that the majority of the teams would be back on independent status, also in effect banned special bodies, limiting Daytonas, SuperBirds, Talladegas and Spoiler II's to 305 cubic inches.

With no more factory edicts on what to run, racing teams sticking with Fords generally chose the 1969 Mercury Cyclone to race for the 1971 season. Some also tried 1970-71 conventional Torino hardtops, which were found to be more aerodynamic than the ill-fated SportsRoof from those years.

Spurred by Bobby Allison's performances in a Holman & Moody Cyclone, Mercury took ten Grand National wins in 1971, its highest total ever. Fords only took four. (The NASCAR record book is wrong, mixing up a June 6 win at Dover, Delaware, by Allison who drove a Ford, while NASCAR called it a Mercury.)

After the 1971 season, the 1969's were too old to run. The 1972 models, with coil rear springs were little enticement to the remaining Ford teams, so the 1971 Mercury Cyclone was used. It was identical to the 1970 models in configuration. The Wood Brothers discovered it was one of the greatest superspeedway cars of all times. After a few driver changes, Pearson got into the car and the Cyclone took nine big track wins in 1972 and ten in 1973. Counting a short-track win in 1973, the total of eleven victories was an all-time high for Mercury.

USAC had two strong Ford teams for 1970, the Jack Bowsher effort with Bowsher and Foyt for drivers and Al Unser driving a 1970 Torino for Rudy Hoerr. Independent Dick Trickle also ran a strong Ford. Fords took eight wins to seven for Plymouth and five for Dodge and won the manufacturer's award. Roger McCluskey repeated as champion, driving a Nelson Plymouth.

Fords used by the Bowsher team included 1969 Cobras and Talladegas and 1970 SportsRoofs. After the 1970 season, Ford also pulled its help from Bowsher and Hoerr. Bowsher made the most of his equipment and continued to field Fords in USAC well into the 1970's.

Ford's drag racing efforts were concentrated in the new Pro-Stock class, which started with pony cars and got down to subcompacts. While individually successful efforts with Torinos were made in 1970 and succeeding years, the days of factory cars, special models and major confrontations were history. The Torino was all grown up and was nearly the size of full-sized Fords of not too many years before. The action was downstream in the smaller cars.

While the 1970 Torino may not have successfully completed its assigned tasks on racetracks, one thing was clear, it was the most popular Ford intermediate ever. Production reached a new high of 407,493 for an eleven-percent increase. The total included the 1970½ Falcon models, which accounted for 56,514 orders (greater than the increase of 40,582 over the 1969 models).

Most popular of the 1970 Torinos was the GT SportsRoof with 56,819 being made. Cobra production came to 7,645.

1971

The curtailment of most racing came to light well after the 1971 models were on the street. The full effect of the antiperformance moves within Ford wouldn't be felt until the 1972 models came out.

General Motors shocked the performance world with the news that all of its 1971 models would run on low-lead or no-lead fuel. That meant cuts in compression and detuning, resulting in less performance. Chrysler cut some of its lesser engines, but Ford said it would get around to taming its engines later and started the 1971 model year pretty much as it had ended 1970.

For once, all the engine sizes remained the same, but outputs changed. The 250 six was rated at 145 hp, the 302 V-8 at 210 hp and the 351 Windsor two-barrel at 240 hp. Fifteen horses were cut from the 351 Cleveland four-barrel, now rated at 285 hp, thanks in part to a compression cut, from 11 to 10.7:1. It still came with Ram-Air with no ratings change.

Back again were the 429's, now called CJ instead of Cobra Jets. The 360-hp mild version was gone, but both 370's were back, with and without Ram-Air. The scooped version got the CJ-R designation. No longer listed was the Super Cobra Jet, but the engine was installed in some 1971 models.

A close inspection of these two Jack Bowsher 1971 Torino SportsRoofs will reveal that the car in the background has the rear window flattened out, while the #2 car has the stock concaved rear glass.

A new divided grille and new rocker panel trim were the most visible differences between the 1970 and 1971 Torino GT SportsRoofs. The 1971 model still had the simulated hood scoop standard.

Model designations were shifted with the Falcon and Fairlane 500 names retired. Torino was the base series, with Torino 500 the next step up. Models were the same as Fairlane 500 and Torino series, respectively, the year before. Remaining the same were Torino Brougham, Cobra and GT series.

Trim changed slightly with a split grille on Torino 500, GT and Brougham models. The GT convertible would be the last for a Ford intermediate. Cobra lost its standard 429 and now came with the 351 4V as standard. The rest of its goodies were pretty much intact. Price was relatively unchanged at $3,295.

Interest in performance cars was on the downswing, as demonstrated by the 1971 Cobra production total of 3,054. Torino popularity in general decreased in 1971's model run with production falling nearly twenty percent to 326,463. New car production only fell five percent for the model year.

1972

With its two-year styling cycle elapsed, it was time for a new Torino for the 1972 model year. This time there was more than a change of sheetmetal. The 1972 Torino was the most changed car in the history of Ford intermediates.

Unitized construction was abandoned in favor of separate body and frame design. High coil front springs were replaced by conventional coils between the control arms. The rear leaf springs departed in favor of coils with four links. Following the GM lead, two-doors went on a shorter 114-inch wheelbase and four-doors got a 118 stretch.

Different grille styling for different series produced a variety of length figures. The base Torino models were 3.6 inches shorter than the fancier Gran Torinos. Two-door Gran Torinos measured 207.3 inches, four-door sedans 211.3 and wagons 215.1.

Width ballooned out to 79.3 inches for most models. Tread grew and was now 62.8 inches for the front and 62.9 for the rear. Heights also varied, but hardtops were just under fifty-two inches, or somewhat taller than the 1971 SportsRoofs and lower than the 1971 regular hardtops.

Though it wasn't known at the time of introduction, the 1970-71 Mercury Cyclone hardtop turned out to be one of the best superspeedway cars of all times. David Pearson dominated the 1972 and 1973 seasons in The Wood Brothers #21 Cyclone.

Torino for 1972 was all-new from road to roof. It had new chassis, body and interior design. The Gran Torino Sport SportsRoof was the sportiest version with fastback styling and a simulated hood scoop. Magnum 500 wheels were added to this example.

More dramatic was the increase in weight, climbing 300-350 pounds. Quietness, comfort and solidness were now valued more than impressive power-to-weight figures. And styling bespoke bulk instead of sportiness. Top-line models got a fish-mouth grille, flanked by quad headlights mounted in the bodywork. Base models received a full-width grille with the horizontal quad headlights mounted in it. Left over from the performance days, rear quarters did bulge out. A straight-bar rear bumper housed the taillights at the ends. Front and rear valance panels tucked under the bumpers.

Greatly simplified was the model lineup. The base Torino series came in two-door notchback hardtop, four-door sedan and four-door wagon. For 1972 Torino four-doors had center posts, but doors were of frameless design.

Gran Torino had the same models as Torino, but also had an extra wagon, the Squire. A Brougham interior was an option. The only hint of performance leanings came in the Gran Torino Sport series, which consisted of notchback hardtop and SportsRoof fastback hardtop. There would be no 1972 Torino convertibles. Side trim was similar to the Gran Torino, but Sport models came with a simulated hood scoop. Insides were similar to Gran Torino models with buckets and console an option.

Debates on the aerodynamic qualities of the fastback were not forthcoming, for few teams would be racing the 1972 Torinos, for reasons stated before.

Standard on the Sport models was the 302 V-8.

It wasn't as if Ford totally ignored its performance past, for the Rallye Equipment Group Option was offered. At a cost of $453, you got the 351 4V, four-speed box with Hurst shifter, competition suspension, G70x14 raised-letter tires on six-inch rims and full instrumentation including a tachometer.

You'll notice thus far horsepower readings haven't been mentioned. Starting with the 1972 models the domestic industry switched to net figures from the previous gross output system. It was just as well, as the reduced compression ratios and tuning cut the output, in some cases to embarrassing proportions.

There was only one true performance engine for the 1972 Torino, the 351 4V with a net horsepower rating of 248 at 5400. The Cleveland engine at first was optional for all models, but later was restricted to two-door hardtops. It was the only 1972 Torino engine available with a four-speed transmission. While 248 may not sound like much, compare that to the 429, which only came with a single four-barrel and rating of 205 hp.

Other engines and their new ratings were the 250 six at 95 hp; the 302 V-8, which was standard for Sport models, at 140 hp; the 351-W at 161 hp; and a new 400-cube V-8, which carried a 168-hp rating. Actual ratings varied a few horses either way; for Ford said that the output of an engine would change depending on models and equipment, to further confuse things.

The 400 used the Cleveland block with a four-inch stroke and one-inch higher deck. Heads had small valves and few parts interchanged between it and the 351's. It was not a performance engine, but rather aimed at sedate motoring with fair economy.

About the only racing significance to the 1972 Torinos was a project by Bud Moore to make the 351 a winner, even though he had to race against 429's and Hemis. It took awhile, but he finally succeeded. In later years when NASCAR went to smaller engines calling for small-blocks, Moore was ready.

There weren't a lot of 1972 Torinos on racetracks around the country, but that didn't mean that properly set up they were not competitive. Leading a pack of cars at Michigan International Raceway in a 1974 NASCAR Grand National race is Buddy Baker in the 1972 Torino SportsRoof of Bud Moore. Moore did pioneer work in developing the 351 Cleveland for NASCAR.

Mercury also redesigned its Montego for 1972 in similar fashion to the Torino. Its fastback was the Montego GT. It too was a rare sight on racetracks until the 1974 season when the leaf-sprung 1971 models became too old to race.

It would be nice to say that the lack of performance models, engines and even convertibles hurt Torino sales and taught Ford a lesson, but such was not the case. The 1972 Torinos were the most popular Torinos of all time, with 496,645 being built, an increase of more than fifty-two percent.

The public was in a luxury mood; all the other competition was warmed over and Torino came in right on target. Gran Torino hardtops were the most wanted with 132,285 being ordered. Most popular of the Sport series was the SportsRoof at 60,974, compared to the notchback's 31,239.

1973-1979

The last Torino we'll glance at is the 1973 model, not for what was new but for what was left. It was the last year for the SportsRoof and four-speed transmission. All performance programs in Ford died during the model year. The so-called energy crisis that followed the model year and cut big car sales put any return to performance on a back burner for years to come.

Styling for 1973 wasn't as clean as the 1972's, as federal bumper standards dictated a rather abrupt grille. Gone was the simulated hood scoop for the Gran Torino Sport.

The public didn't mind. New car production rose 15.5 percent. Torino, which had stiff competition from the all-new GM intermediates, held its own. Production stayed virtually the same at 496,581. SportsRoof hardtops easily outpaced the notchbacks in the Gran Torino Sport series, 51,853 to 17,090, but were discontinued at year's end.

Torino production continued through the 1976 model year. Restyling took place for the 1977 models with the Thunderbird now built on the two-door chassis. Regular Torinos were renamed LTD II and remained in production through the 1979 model year, after which they were not directly replaced.

After the 1971 models, Torinos ceased to be all-out performance cars, but few really cared. Times had changed, the market had changed and it was a rare Torino buyer who even knew what a Thunderbolt was.

Last of the SportsRoofs was the 1973 Gran Torino Sport. Despite being more popular than the Gran Torino Sport notchback, it was dropped at the end of the model year. Optional Magnum 500 wheels were fitted to this model. Gone was the simulated hood scoop.

After the 1971 Mercury Cyclone was too old for NASCAR in 1974, The Wood Brothers switched over to a 1973 Montego fastback for the 1974 and 1975 NASCAR Grand National seasons. David Pearson did the driving, but the new car could not match the track record of the old.

Pro-Stock Conquests

One of the most popular forms of professional drag racing in the 1970's was Pro-Stock competition.

It started in 1970 with the idea of a stock-looking body and a fairly stock chassis, powered by a highly modified engine of the same make as the body. Big name drivers were attracted and the fans loved it.

At first there were mainly pony and compact cars, like the Ford Maverick and Mustang, Chevrolet Camaro, Plymouth Barracuda and Valiant Duster and Dodge Dart and Challenger.

Like any other form of racing, it didn't take long for things to get more sophisticated and the rules to be bent. When the subcompacts came out for 1971, Vega and Pinto pro-stocks followed. Chassis modifications were allowed and tube frames came on stream.

Despite the changes, the pro-stocks still weren't as radical as the funny cars and made themselves a nice niche between the funnys and the stocks. As the rules changed over the years, various Ford powerplants came into favor including the 351 Cleveland, the sohc 427 and Boss 429.

Ford's involvement in drag racing, especially pro-stock, lasted well into the 1970's, long after other forms of motorsport were abandoned. Indeed, the rules were often kind to the Dearborn machines.

Pro-stock's timing couldn't have been better. In the early 1960's, super stock racing was very popular, as it reflected the performance available from the new cars and provided a showcase for the endless stream of new muscle cars from Detroit.

In 1970, that flow was about to slow to a trickle. In the coming years, there would be fewer high-performance engines and models in which to install them.

Since pro-stock wasn't tied to the latest cars, but only the latest bodies, it was insulated from changes in corporate performance policy that eventually turned super stock into a class for old cars.

Though Ford was involved in drag racing, the relationship of the drag strip to the production line was no longer there. Pintos may have been used in pro-stock, but no high-performance Pinto evolved. The same can be said with Maverick. The market and Ford's product philosophy could no longer accept such a car.

In the 1980's, Ford enjoyed a resurgence in Pro-Stock competition. Led by long-time racer Bob Glidden, Ford has enjoyed many good seasons in Pro-Stock. Glidden has used several bodies for his racers, including Fairmonts and his latest Thunderbirds.

Here Bob Glidden is shown in the 1973 NHRA Nationals at Indianapolis Raceway Park. His 1972 Pinto was top qualifier at 9.033 sec/152.54 mph and beat Wayne Gapp in the final run for the win.

Ford Wins Indy

Of all the projects that Ford got into in its Total Performance era, the most daring was the building of an engine to compete in the Indianapolis 500. Domestic passenger car builders could get away with a lot of things in racing, but spitting into the face of the legendary Offenhauser at Indy wasn't supposed to be one of them.

The four-cylinder strictly-for-racing powerplant had dominated the Brickyard since the 1930's, and no matter how strong engines like the early Chrysler Hemi or Chevy small-block were on the street and in other forms of racing, they were unable to make even a scratch in the facade of the Indy establishment.

Ford set about to change that by using its Fairlane V-8 as a basis. In stock form the 289 was just too heavy, despite having one of the lightest cast-iron blocks in the industry. Ford engineers cast it in aluminum and made several changes, getting its weight down to 350 pounds and its horsepower up to that number.

The engine would not be enough to do the job in the big roadsters that were the rule at Indy, but in a lightweight, rear-engine car, similar to those in Formula One competition, they would be very competitive.

That brought in the other component of Ford's Indy program, Colin Chapman, president of Lotus Cars Limited of England, which built and raced Formula One cars. Together Ford and Chapman came up with the Lotus-Ford Indy car, which was entered in the 1963 Indy 500 with drivers Jim Clark and Dan Gurney.

Gathering much publicity, the cars lived up to the expectations by qualifying well and running up front. Clark did not win, but finished second to Parnelli Jones in an Offy roadster. Gurney finished seventh. Proving they could win, Clark and Gurney were entered in a two-hundred-miler in August of that year at Milwaukee. Clark ran away from the field and Gurney took third.

Indy's Borg-Warner trophy evaded the Ford effort in 1964 when A. J. Foyt took the last win for an Offy roadster, but now it was only a matter of time.

The reign of the Offy roadster officially ended in the 1965 Indianapolis 500 which, appropriately, went to Clark. Ford power was established and went into the 1970's as a major force in Indy car racing.

Offys were turbocharged and did stage a comeback in rear-engine cars, but Fords got turbocharging too and gave them a run. Eventually Ford turned over the costly production of its Indy engine to Louis Meyer, who was involved in Offy production.

Ironically, both the Offy and Ford engines were replaced in Indy car racing by the Cosworth Ford V-8 from England, which in turbocharged form dominates the sport today. The engine started life in Formula One, just as the Lotus chassis design did.

While the Cosworth V-8 faded from Formula 1 by the mid-1980's, it remains a dominant force in Indy Car racing, though much new competition is on the horizon as 1987 approaches. For Formula 1, Ford has a new Cosworth V-6.

After three years of trying, a Ford-powered car won the Indianapolis 500 in 1965 when Jim Clark crossed the finish line first in Colin Chapman's Ford-powered Lotus Model 38.

Shelby Cobras

Chevrolet spent millions of dollars and countless man-hours to bring the Corvette from a six-cylinder boulevard cruiser to a genuine high-performance sports car. That process had been going on about nine years when Carroll Shelby took a Fairlane V-8 and put it into an AC Ace in 1962.

The AC was a mid-1950's vintage car of somewhat dated design, and the 260 V-8 was a new small-block lightweight powerplant with potential for much performance and compact dimensions that did not upset the AC way of things all that much.

In one move, Shelby sent Chevrolet back to the drawing board. The timing was perfect for Shelby. Ford was interested in a performance sports car, but not tooling-up for their own car. AC had excess capacity and Shelby had the drive to put it all together. A factory was set up in Los Angeles and AC Cobras were put in production mating the 260 to the AC body and chassis. After about seventy-five cars, Ford's 271-hp 289 was used.

This image in the rearview mirror of a Corvette A-Production sports car racer was enough to strike fear into your heart. Bob Johnson cruises along in his 289 AC Cobra in 1964.

When Cobras and Corvettes met on a racetrack, Cobras usually feasted. Corvette, with its all-new Sting Ray body for 1963, was just too heavy. Chevy tried to counter with Grand Sport lightweight versions, but a GM antiperformance edict in 1963 nipped the plan in the bud and only five were made.

The 289 was sufficient for a while, but starting in 1966, Cobras got new bodies and chassis and 427 power, with mighty side-oilers producing up to 480 hp.

All good things have to come to an end, and for Cobra the end was in 1967. The impending federal emission and safety regulations for 1968 were too much for the Cobra. The cost of changing the car and engine to meet them was too high for the volume; and even if the changes were made, the result would have been a lesser performing car than the 1967's.

In all, about 630 small-block and 510 big-block Cobras were made and each one that exists today is extremely valuable and highly sought after by collectors.

Though the original Cobra died in 1967, the name did not. Ford attached it to the 1968 Shelby Mustangs, a 1969 through 1971 performance version of the Torino, several engines and, starting in 1976, a specially trimmed Mustang II.

Ford applied the Cobra name to the new-generation Mustangs as an option package from 1979 through the 1981 model year, but it has been dormant since then, at least through the 1986 model year.

Meanwhile, Carroll Shelby became an adviser and limited-production-car builder with Chrysler Corporation and his old friend Lee Iacocca. The Shelby name has been on versions of the subcompact Dodge Charger and sporty Dodge Daytona, and more is coming.

The AC-Cobra has gained cult status in the kit car area with many copies being made today, with power coming from everything from a Volkswagen Beetle engine to an authentic Ford 427 side-oiler. Even one of the new versions of the Cobra is using the original tooling.

Cobras ruled A-Production SCCA sports car racing, taking the national title every year from 1963 through 1968. Here George Montgomery is shown on his way to victory in a 1965 A-Production race at Road America with a new 427-powered Cobra.

Cobras didn't restrict their winning to sports car races, they also did well in drag events. Ed Hedrick proved that in the 1967 NHRA Nationals at Indianapolis Raceway Park when he drove his 1964 289-powered Cobra to D/Sports honors.

CHAPTER SIX

Phenomena On Wheels Mustang And Cougar

Perhaps a statement from *Mechanix Illustrated*'s Tom McCahill summed up the Mustang phenomenon best: "From the day Ford introduced its cross-bred little horse, it took off faster, saleswise, than the hottest thing the Cape Kennedy kids have been able to muster. Every once in awhile in the automotive business a new model appears from left field and becomes an instant success. This isn't true every year or every five, but when it occurs, it shakes the big domes of the competitive camps right down to their toenails."

Literally from the day it was introduced, April 17, 1964, the Mustang was a runaway success. Volumes and volumes of material have been written on the Mustang, most of it liberally spiced with superlatives.

However, from a performance standpoint, Mustang provides an example that racing and horsepower do not always sell cars. Some of the best years for racing and greatest increases in true engine output coincided with the biggest drops in sales. True, there were other market factors, but the improving muscle car status of Mustang was not enough to offset them.

From the beginning, Mustang was aimed at a variety of potential buyers, from the economy-minded to tire-burners, with several stops in between, all with a sporty flair. Also from the beginning, the image of the Mustang was carefully controlled—from the preproduction show cars to the cars that competed on the road courses and drag strips.

Reasons for Mustang success can be traced to the successful combination of two elements, sports cars and compact cars.

Sports cars like the MG TC and TD became popular after World War II and were joined by domestic examples like the Chevrolet Corvette in 1953 and the Ford Thunderbird a little over a year later. However, only holding two passengers in comfort, they were impractical for a family.

The domestic compacts, like the Rambler American of 1958, Studebaker Lark of 1959 and the Big Three's 1960 Ford Falcon, Chevrolet Corvair and Chrysler Corporation's Valiant, were sporty in size, but for the most part dull in concept.

Chevrolet was most successful at combining the two concepts with its mid-year Corvair Monza 900 coupe in 1960. It had bucket-type front seats, tastefully reserved trim, interesting mechanicals with an air-cooled rear engine, and best of all, it could be serviced down the street at the local Chevy dealer. Corvair sales shot up and imitators followed, like the Falcon Futura.

The Corvair's design was limited to the small-inch six, meaning it would have a hard time competing in the growing performance orientation for the compacts. It got a turbocharger for its Spyder option and a convertible in 1963, but Ford was already at work with an answer that could use its new small V-8, the Mustang. Mustang's performance potential would be used, right from the start, with Falcon and Fairlane parts available to bolt in.

Mustang didn't hit the market cold. An ingenious preintroduction scheme prepared the buyer for the concept and the name. In 1962 the Mustang I show car garnered wide publicity. It was a mid-engine, two-passenger sports car and a break from anything Ford had on the market. Both pure sports car and domestic car fans took notice.

Neat as it was, the Mustang I didn't have the wide market appeal needed. Bringing the Mustang name closer to the final product was the Mustang II show car (which was of four-place configuration) and, except for a chopped-looking top, was close to what would be on the market the next year after its 1963 introduction.

When the production Mustang came on stream on that fine spring day in April of 1964, the public was ready. If ever a car was right on, right out of the box, it was Mustang.

It only came with sporty looks—no plain-Jane model, it came with bucket-type seats standard. Thanks to the use of many parts already in production it could be offered at a low price. And, calling on the performance development of several projects, could be had in versions from Falcon mild to Cobra wild.

Press reaction was unlike anything Ford had seen for years. Iacocca and a bright red Mustang hardtop made the cover of *Newsweek*'s April 20, 1964, issue and just about every automotive magazine that came out near the introduction date.

Praise was nearly universal, criticism was buried. "The Mustang is definitely a sports car, on par in most respects with such undisputed types as the MG-B, Triumph TR-4 or Sunbeam Alpine," said the usually reserved *Road & Track*.

"Unless we miss our guess, the exploding youth market will have the assembly lines working nights and weekends trying to keep up. Ford is in on the ground floor with their Mustang, and we can guarantee this car is going to be popular," noted a prophetic Ray Brock in *Hot Rod*.

"A market which has been looking for a car has it now. Ford Motor Co. has finally recognized that market, the voice of which has been easily detectable amid the pages of automotive enthusiast magazines, and has started a round-up of its state of the Total Performance art to produce the Mustang," reported *Car Life*.

When a car can win praise from the sports car, hot rod and middle-of-the-road automotive press, it has got to be something special—and Mustang certainly was.

Press reaction, though, was nothing compared to buyer response. Early projections hoped for sales of 100,000 of the 1965 models, over an extended model year that ran from the April 1964 introduction to the fall of 1965.

Projection makers were nearly as busy as the doors on the showrooms of Ford dealers. Model-year sales for 1965 came within a sneeze of the half-million mark, with production pegged at 559,451. Mustang became the unexpected third most popular nameplate, behind full-sized Chevrolet and full-sized Ford.

There were lots of cars that came close to the Mustang in various areas, but none that put it all together as well. Suddenly Chevrolet's Corvair, its new Chevy II, Plymouth's Valiant, Dodge's Dart or even Ford's own Falcon were pale by comparison. Only Plymouth's Valiant Barracuda, introduced at the same time as Mustang, came close.

Well, what made the Mustang so great anyway? The answer was a sporty basic car and a long option list permitting the buyer to tailor his or her Mustang to just about any kind of taste, all at a relatively low price.

Let's look at the components of the early 1965 models.

Ford gained considerable experience with unitized body and chassis construction starting with the 1958-model Lincolns and Thunderbirds. Lightweight unit bodies were developed with the Falcon program and expanded upon with the 1962 model introduction of the intermediate Fairlanes.

Long hood, short deck styling on the early Mustangs set an industry trend for many years to come. It wasn't a particularly efficient design, but hundreds of thousands of buyers didn't seem to care, as Ford had a problem keeping up with the demand for early 1965 models, like this two-door hardtop.

Mustang provided the next step with a platform chassis and a box section for the engine compartment. To this the body was welded to form the unit. It was not the same as the Falcon, contrary to rumors.

However, there were similarities. Front suspension was the high coil, shock tower arrangement used on Falcon and Fairlane, while semi-elliptic leafs located the rear axle.

Styling resembled the Mustang II show car in concept. Two-door hardtops had a higher, formal roof with small quarter windows, while convertibles followed the lines with a manually operated top standard. The long hood, short deck theme would also take the domestic industry by storm. Thin bumpers, generous splash or valance pans and a well done wide, low grille with the Mustang horse inside and single headlights outside of it, made the front look much lower than it was. A sculptured concave indentation on the side was tastefully trimmed, letting the sheetmetal show off its just-right contours.

At the rear, the same visual treatment as the front was used with shallow bumpers and deep splash pans. Simple three-lens taillights held down the outer ends of the panel between the trunk and bumper.

Inside there wasn't really anything unusual, but coupled with the fact that bucket seats and carpeting were standard and it came in a sporty outside package, it put the icing on the cake, so to speak. Twin coves marked the padded dashboard, with a Falcon-based instrument cluster on the driver's side and similar-shaped glove compartment for the passenger. Rather plain door panels were color-coordinated with the buckets and bench rear seat.

Dimension-wise, the package was compared to both the Falcon and two-seater Thunderbird. Wheelbase at 108 inches for Mustang was 1.5 inches less than Falcon and six inches longer than T-Bird. Length of Mustang and Falcon came out identical at 181.6 inches and 0.2 inch longer than the Bird. Tread width of fifty-six inches matched the Thunderbird and was within a fraction of an inch of Falcon, depending on model. Mustangs were 68.2 inches wide, 3.4 less than Falcon and 4.6 under a 1957 Thunderbird.

Weights depended on many things, but starting in the 2,700 range, came in some 400 pounds under the early Bird.

Just like many other facets, the powertrain situation for Mustang didn't offer anything unique, but in combination with other options they made the Mustang a car that spanned the range of buyer interest.

Basic powerplants for the Mustang came right out of the Falcon. Standard for both hardtop and convertible models was the 170-cube 101-hp Falcon six. The next step up was the 164-hp 260 V-8, which started life in 1962½ Fairlanes and became available in 1963½ Falcons.

A new version of the 289 V-8, with four-barrel carb and 9:1 compression (which meant regular-grade gasoline) advertised at 210 hp, held middle ground on the

Early competition plans for the Mustang included following up on the Falcon's successful rally record, in the U.S. and Europe. Things got off to a good start as Peter Proctor and Peter Harper finished first and second in their class in the 1964 Tour de France. After 1964, however, serious rally efforts for the Mustang took a back seat at Ford to other Total Performance projects.

With sporty bucket-type seats standard, Mustang buyers in 1964 got a neat package no matter how much, or little, they paid. This early 1965-model convertible has the optional console, Rally-Pac on the steering column and four-speed transmission. The instrument cluster was Falcon-based.

option list and was the last step before the top offering, the 271-hp, solid-lifter High Performance 289 V-8.

The latter version was used to light the enthusiast's fire. It had served time under the hood of Shelby Cobras and assorted Fairlanes and was capable of great things. *Hot Rod* nailed the go pedal on one with a four-speed and 4.11:1 Equa-Lock gears; it zipped to sixty in 6.9 seconds and completed the quarter in 15.5 seconds. *Road & Track* tested a high performance 289-equipped early Mustang with heavy-duty suspension after testing earlier standard models and noted, "The effect is to eliminate the wallow we experienced in previous Mustangs and to tie the car to the road much more firmly."

Like other features on the early Mustangs, engine availability would change during the extended model year, with major alterations taking place when the traditional 1965 models came out in fall of 1964. Some Mustang fans like to argue over whether the early Mustangs were 1964 or 1965 models, but Ford chose to register them as 1965's, so the best way to look at the situation is that they are all 1965 models and there were mid-year changes.

Transmission availability at the start included a three-speed manual box standard on the six, with the four-speed English Ford unit and three-speed Cruise-O-Matic optional. The 260 V-8 came with three-speed all-synchro as standard and Cruise-O-Matic optional, though some sources report the 260 only came bolted to the automatic.

Mild 289-transmission choices included those for the 260 and added a four-speed manual transmission as an option. The 289 High Performance V-8 came bolted to a four-speed and also came with heavy-duty suspension. All transmission shifters were on the floor, another feature which added to the sporty image.

Axle ratios for non-high-performance engines varied between 2.8 and 3.5:1. For the HP, you could get a 3.5 standard and options of 3.89 and 4.11 gearing.

Standard tires were 6.50x13 on 4.5-inch rims with four bolts for sixes and five bolts for V-8's. Optional, but standard with air conditioning, were 7.00x13 units, while cars with the handling package got 6.50x14 rubber on five-inch rims. An early option included 5.90x15 Firestone Super Sport tires on five-inch rims.

Brakes started out very Falcon-like with nine-inch drums all around for the six and ten-inch drums for the V-8. The small drums and absence of discs on the 289 HP brought groans from the motoring press and were cited as one of the few flaws in the car. Though similar in design, sixes got Falcon front suspensions and V-8's got heavier Fairlane-based units.

If no other fact was impressed upon you upon reading or seeing Mustang ads in 1964, it was the low base price of $2,368 f.o.b. Detroit, as they said, even though Ford was in Dearborn. However, low initial price was only part of the program. Who could resist equipping the Mustang with a number of desirable options? Not many.

Popular early options included, in addition to the powertrain items already mentioned, air conditioning, with outlets mounted under the instrument panel;

When Ford introduced the rest of its 1965 models, in the fall of 1964, it added the 2+2 to the Mustang line and made several powertrain changes. Functional rear quarter vents, a sloped roof and a large rear window set the 2+2 off from the conventional hardtop.

console between the seats, Rally Pac (a tachometer and clock package that straddled the steering column); wire wheel covers (for fourteen-inch wheels); accent stripes and rocker panel moldings; power roof for convertibles; and all the usual audio, convenience and appearance items.

The second phase of the 1965 Mustang model year started with the introduction of the rest of the 1965 Ford line. Though the basic design and styling remained, there were many detail changes, a new model (the 2+2) and new engines. It also wouldn't be the final changes; more were made in spring of 1965 for the one-year anniversary.

Most spectacular, visually, of the fall 1964 changes was the addition of a third model, the Fastback 2+2 two-door hardtop. It shared sedan and convertible styling from the beltline down, but had a roof of its own with single side windows for the doors and functional air louvers styled into the roof quarters behind them. The top sloped from about the front-seat-passengers' head room position to the shortened trunk lid. The louvers were adjustable from the inside. The rear seat could be folded down to form a flat storage area, and there was a trunk access panel behind it for storing long items. It was an attractive package with a base price of $2,589.

Engine choices got a going over, with the 170 six being dropped for a redesigned 200-ci six with seven main bearings and a rating of 120 hp. It was standard on all models. Also axed was the 260 V-8, which was replaced by a 200-hp, two-barrel version of the 289. The four-barrel 289 got a boost to a 10:1 compression ratio and new rating of 225 hp at 4800 rpm.

Both the 200- and 225-hp 289's came with an all-synchro three-speed gearbox standard and the option of Cruise-O-Matic or a four-speed.

A much needed option, ventilated disc front brakes, joined the list with the other items. Also added were wider 6.95x14 tires, which replaced the smaller 6.50x13 units as the standard for the regular V-8's. Standard for the High Performance V-8 were dual red-stripe, premium nylon 6.95x14 tires. The narrow 5.90x15's were dropped. Styled steel wheels were added to the option list.

A number of other small detail changes were made at the regular 1965 introduction, like making the front passenger seat adjustable, affixing inner door handles and window winders differently and changing the length of the nameplate on the side. But the next big round of changes was saved for Mustang's first birthday in April of 1965.

No new models or engines came on board, but an impressive option package, the GT Equipment Group, was now available on all three models. Visible outside differences included a modified grille with a pair of rally-type fog lamps mounted in the revised grille bar, GT badges on the front fenders with stripes parallel to and just above the rocker panel, the individual letters in the middle of the stripes on the front fender spelling Mustang and chrome-plated exhaust extensions that were attached to dual exhausts and ran right through the splash pans, thanks to the omission of bumper guards.

Inside, the most noticeable part of the GT package was a new instrument cluster of five round dials including a larger 140-mph speedometer, and four smaller gauges for fuel, oil pressure, amps and temperature. The gauges were set in a black, crackle-finish oval which matched the glove compartment door surface.

To motivate the gauges, the standard engine was the 225-hp 289, now called the Challenger Special V-8. The 271-hp engine was optional.

Completing the GT option underneath was a heavy-duty suspension, disc front brakes and quicker 22:1 steering ratio.

Two other notable options were added to the Mustang list in the same general time period, a bench front seat and the Interior Decor Group. The bench seat, the first for Mustang, had a fold-down center armrest and was available in hardtop and convertible models.

While the GT option had a sporty theme, the Interior Decor Group added a Western flavor. Available in all three models, it featured upgraded seat designs with embossed horses on the back surface of the front seats, wood-grained and chrome steering wheel, walnut-toned glove compartment and surround for the GT-like instrument cluster (and console when ordered as an option), courtesy lights in the doors and other details.

Dick Brannan was a member of the elite team of factory-backed A/FX Mustang drivers in the 1965 drag racing season. The cars featured lightweight bodies, bubbled hoods, sohc 427 power and special front suspension.

The combination of sporty and luxury options being introduced at the same time again held an appeal for a variety of tastes. Needless to say, the spring announcements were timed to keep the fires going by adding to the attractiveness of the Mustang.

Also keeping the interest up was a somewhat limited program of racing for the Mustang. Considering Ford's other endeavors, and the potential appeal of the Mustang to younger buyers attuned to racing, it was rather sedate. However, the Shelby Mustang was to carry the racing banner as well.

An early component of the Mustang's motorsports exposure program wasn't as a race car but rather as pace car for the 1964 Indianapolis 500 in May. Though the race had a fire that killed two drivers, Eddie Sachs and Dave MacDonald, it also resulted in Mustang appearing before the largest crowd there was in racing. In one form or another, Mustangs have been appearing before racing crowds ever since.

Initially, two types of racing would be chosen for Mustang showcases, rally competition and drag racing. (Sports car production class racing would be left to the Shelby.) Ford made a very successful entrance into the rugged sport of European rallying with its 1963 and 1964 Falcon Sprints. Mustang was a natural mid-season successor to continue Ford's inroads against traditional European cars.

Success, however, was limited with the 1964 Tour de France being the only major win, which was taken in the touring class by the team of Peter Proctor and Andrew Cowan in a Mustang hardtop.

Budget cuts for 1965 eliminated the extensive program of past years, though some Mustangs were entered by Ford's European branches and private individuals. Mustangs became instant favorites with private individuals in the U.S. and Canada from introduction on, but major factory rally support was several years off.

Far more extensive was a drag racing program for Mustang. During the 1964 season Ford built and sold 427-powered Fairlane Thunderbolts and did quite well in the super stock arena. It also experimented with a few 427 Falcons in factory experimental competition.

Technology was progressing rapidly with the biggest engines being stuffed into the smallest cars. Mustang wouldn't have to wait long to get involved.

A few private Mustangs with 427's shoehorned-in made the strips in 1964. One was the Quarter Horse driven by Ron Pellegrini with a fiberglass front clip and 427 High Riser stuffed between rather stock front end shock towers.

However, the factory got involved for 1965. A change in NHRA's super stock requirements called for one hundred examples to be built instead of fifty the previous year. The bean counters at Ford said no and decided Mustang drag cars would compete in the factory experimental class, where production numbers weren't needed.

While the overhead-valve 427 High Riser got the job done in 1964, the more potent single overhead cam (sohc) 427 was ready for duty in 1965. It was the engine Ford was trying to get NASCAR to legalize for stock car racing. Drag racing was an ideal proving ground, even if it wasn't the first choice.

The sohc Mustangs went into A/FX, the top class. Ford also entered sohc 427 Galaxies in B/FX and 289 Galaxies in C/FX. Chosen as a basis for the Mustang A/FX was the 2+2. A fiberglass front clip including a bubbled hood and fiberglass front door panels further cut the weight.

With the sohc wider than the regular 427, which was a tight squeeze in a wider Fairlane, something had to be done to the Mustang front end to get the big mill to fit. A system of using leaf springs as torsion bars, attached to the lower control arm allowed the shock towers to be omitted and the big engine to fit nicely. Traction bars delivered the torque in the rear.

Wheelbase was altered two inches, within NHRA's rules and ten-inch-wide tires fit under the wheel wells, thanks to a narrowed axle. The engines had a pair of four-barrel carburetors, but lacked the elaborate air-ducting systems that the Thunderbolts had.

Unlike the Thunderbolts, the A/FX Mustangs were not available through Ford dealers, but rather only to racers. They were built on a special line at Holman & Moody in Charlotte, North Carolina.

In the past, big Ford and Thunderbolt drivers had their hands full with intermediate-sized Plymouths and Dodges. The tiny Mustang would end up taking care of the MoPar problem pretty well, but there were still the A/FX Comets entered by Lincoln-Mercury Division, which also got the sohc engine for 1965.

As usual, the first major test came at the Los Angeles County Fairgrounds at Pomona, California, where the NHRA Winternationals were held in February. Factory stock eliminator turned out also to be the A/FX competition, due to a rain-out the day before, pitting the Mustangs, Comets and MoPars in one fierce pack.

Five Mustangs were in the first round, driven by Gas Ronda, Len Richter, Bill Lawton, Dick Brannan and Phil Bonner. After the first round, four were left, with only Brannan losing while three of the four Comets went down as did two of the four MoPars.

In the second round Bonner shut down and dropped out and Richter beat Ronda, leaving Richter and Lawton in the next round with the Comet of George DeLorean and Plymouth of Tommy Grove.

Lawton and Richter both won, but Richter did not make the call for the final thanks to a twisted axle, so Lawton took the bye and trophy in the Tasca Ford Mustang.

Mustangs did well in A/FX with the season highlight at the NHRA Nationals at Indianapolis Raceway Park. The final round boiled down to a run between two long-time Ford competitors, Gas Ronda and Les Ritchey. Ritchey's Performance Associates Mustang had Weber carbs atop his sohc and though he had a slower run, 10.67 seconds to 10.63 for Ronda, Ritchey got to the end of the quarter first and won the class.

Another application of the Mustang in drag racing was as a funny car. Elongated hoods and fuel injection, later supplemented by supercharging, made these cars faster than the A/FX cars. But since they had little application to the production Mustangs, much less than to the A/FX cars, they will not be covered in detail here.

Yet another part of the Mustang performance story was the Shelby version. Carroll Shelby put a Fairlane 260 V-8 into an AC Ace, a two-seat, rather dated British sports car in 1962 and with the blessings of the Ford Motor Company, turned it into a Corvette-beater in sports car competition. Shelby did development work on the Ford small-block, eventually putting 289's into Cobras. Later he followed with 427's.

With this background, it was logical that Ford turned to Shelby to see what he could do with the Mustang, especially for production sports car racing, which was Shelby's long suit. The result was the 1965 GT 350, a 2+2, which Shelby took from the Ford assembly line at San Jose, California, and completed in his factory at the Los Angeles International Airport. Since we're primarily interested in the regular Mustang, only their major points will be covered. They got heavy-duty suspension with Koni shocks, a highly tuned 289 with single four-barrel, tubular headers and rating of 306 hp at 6000 rpm.

A fiberglass hood with scoop, no back seat and four-speed transmission were also in the package. Cars were painted Wimbleton white and got blue racing stripes above the rocker panels. Wide blue stripes were put on hoods, roofs and trunks by dealers, and a few went on at the factory.

A few GT 350R models were built and these were strictly race cars, being legal for B-Production in SCCA. To prove it, Jerry Titus won the national title for the class in a 1965 GT 350R.

1966

How do you follow up an act like the 1965 Mustang? The answer for 1966 was with more of the same, as Ford placed Mustang on a two-year styling cycle, with its change years alternating with the intermediate Fairlane.

Only trim changes were made to the basic body. There were no model or engine changes. Background material for the grilles changed from a honeycomb stamping to an egg-crate design of extruded aluminum with the horizontal bars in bright trim. Only the center emblem was mounted in front of the bars. On GT-optioned models, the bars and lights of 1965 returned and the background was flat-black.

Minor changes marked the 1966 Mustang. This convertible shows off its new grille background and windsplits on the rear quarter panel. Among the options on this example are styled steel wheels and a 289 V-8, with emblem on the front of the front fender.

A chrome piece with three windsplits was added to the rear of the concave area on regular hardtop and convertible models, making it appear like a scoop. Wheel-cover design also changed, to give the 1966's a slightly different appearance.

Inside, all Mustangs now got the five-dial instrument cluster, including the 140-mph speedometer—even on six-cylinder cars.

Mechanically there were detail improvements, one of which was the availability of Cruise-O-Matic with the 271-hp V-8.

Mid-year wasn't excitement time for 1966, as only a Sprint 200 option was announced for the Mustang hardtop, and only involved special trim items and a chrome air cleaner for the standard six for underhood attention. It was part of Ford's celebration for building a million Mustangs.

Actually, there wasn't a drastic need for more gingerbread, as Mustangs were selling like proverbial hotcakes. Model year production reached a record 607,568, up from the 559,451 produced the previous model year.

Still, there was no major competition for the Mustang during the 1966 model year.

In the racing world, Mustang found a new home. SCCA planned a series of races for Group Two sedans at road courses across the country. The Trans-American Championship (or Trans-Am as it became known) was for cars under two liters (B-sedan) and two-to-five liters (A-sedan) with a limit counting allowance for clean-up of 305 cubic inches. Several modifications of car and engine were legal. This would seem a natural for the Shelby GT-350, but for some reason, the notchback hardtop was the only Mustang homologated, and that was the model to be raced.

At first, there was no factory backing from Ford for the series. There was some backing from Chrysler for rally driver Scott Harvey in a Team Starfish Barracuda and possibly some for Bob Tullius in a Dodge Dart hardtop. Mustangs were entered by independents.

Thanks to the under-two-liter class, races were not runaways for the American cars, as Alfa Romeo GTA's and Lotus Ford Cortinas chased the larger sedans all over the place. When the season ended, Ford won the Manufacturer's Championship for the big cars with forty-six points for Mustang, compared to thirty-nine for Plymouth and thirty-three for Dodge.

Points were only given to the make of car, not the driver. Tullius opened the season with a win at Sebring, Florida, in his Dart, but actually finished second to an Alfa in the race driven by Jochen Rindt. The independent team of Bob Johnson and Tom Yaeger won at Mid-America Raceway in Wentzville, Missouri, and again at Virginia International Raceway, but their wins were separated by a Barracuda flag at Bryar Park, New Hampshire.

Mustang held a one-point lead going into the twelve-hour enduro at Marlboro, Maryland, but Tullius and Tony Adamowicz teamed to give Dart its second win.

Ford realized it might have a chance of winning the series and entered a factory car at the next event. The Shelby-ized Mustang hardtop was entered in the six-hour event at the Green Valley Raceway near Ft. Worth, Texas, for Don Pike and John Timanus. However, independents John McComb and Brad Brooker won for Ford; and going into the final event at Riverside, California, Ford and Plymouth were tied with thirty-seven points each.

Sports Car Graphic Editor Jerry Titus drove the factory-supported Mustang at Riverside, won the race and the title for Ford and put the car on the cover of the December 1966 issue of his magazine.

Newly designed wheel covers were standard on the 1966 Mustang, to go along with the new grille and trim on the rear quarter panel. This hardtop had the newly standard rocker panel trim and a 289 V-8.

Mustangs were back in the A/FX biz in drag racing for 1966. *Car Life* looked at one of the new models and found, with its 5.14:1 gears, 10.50x15 slicks and engine rated at 600 hp at 7200 rpm, it was good for an 11.7-second quarter, with a trap speed of 128.6 mph. The magazine estimated it hit sixty from a start in less than four seconds and noted the sohc could be wound up to 8700 rpm with relative ease. Weight just nudged past the lower class limit of 3,200 pounds.

Pomona was ready for the 1966 Winternationals and the Mustangs were ready for Pomona—or so they thought. The A/FX final was a run between Les Ritchey in the Performance Associates Mustang and Jerry Harvey in the Bob Ford Mustang. Ritchey blew a transmission the day before and, lacking a new one, cobbled one together out of spare parts. It lasted three gears and let go, giving Harvey the class win.

A new system put FX winners into the street eliminator runoffs, while super stocks and down went into stock eliminator finals.

Harvey found himself in a group with everything from modified Volkswagens to Cobras. He had planned to use a specially built engine for the Winternationals; but since it blew in a meet before, he had to be satisfied with a near-stock unit. It seemed to be working well as he blew away a Plymouth, Dodge and Corvette on his way to a final-round confrontation with the very strong B/FX 1966 Galaxie of Mike Schmitt.

Harvey waxed the big Ford but good with a 10.68-second run to Schmitt's handicapped 11.51 and took the street eliminator title. "I can't believe it. The most important race of my life, the biggest one I've ever seen and I won!"

A/FX cars lost some of their appeal in 1966 as the funny car attracted the attention of the fans and some of the top Ford drivers like Bill Lawton and Hubert Platt. Lawton topped the class for them, A/XS, at the NHRA Nationals on Labor Day. The XS means experimental stock.

Back again for 1966 were the Shelby Mustangs. The basic package remained the same, but the first signs of watering-down to tastes of the average buyer could already be seen. Early 1966 GT-350's were leftover 1965 models, so they retained the 1966 mechanical features including the heavy-duty suspension and Detroit Locker differential. Koni shocks and the Lockers were optional after that. A special panel on top of the dash for tachometer and oil-pressure gauge was eliminated and a rear seat was available, with the spare back in the conventional location, after being placed inside the car the previous year.

Outside, the most visible features were the replacement of the rear quarter louvers in the roof with plexiglass units and the installation of a side scoop in the rear quarter panels. There were other trim changes as well. Five colors were available for 1966 compared to one the year before.

An option in place of the Borg-Warner T-10 close-ratio four-speed was the C-4 Cruise-O-Matic. Late in the model year, a Paxton supercharger was made optional for the GT-350.

An unusual deal helped Shelby production and resulted in many of the 1966 models being black. Hertz Rent-A-Car ordered 936 GT 350's for its rental customers. While the amount may not seem high, it should be remembered that the entire 1965 model run had come to only 562.

Early Hertz cars had four-speed boxes, but most came with the C-4. Most were also black with gold stripes in the usual places and designated GT 350-H. More than one was rented and entered in weekend races of several types. Hertz did not repeat the special order for 1967. Later Hertz did buy some more Shelbys, but they lacked the special identification of the 1966 models.

"These cars are available to business travelers who want a change of pace in motoring, by sports car owners away from home and by vacationers who con-

B-Production, long the mainstay of smaller-engine Corvettes, was invaded by the Shelby GT 350 in 1965, with the modified Mustangs winning the national SCCA class title in 1965, 1966 and 1967. Shown in a 1966 event at Road America are Dr. David Ott (#1-1961 Corvette) and David Pabst, Jr. (#17-1965 Shelby GT 350).

sider driving an enjoyable holiday sport," said the brochure on the GT 350-H. A special rental agreement was required.

The Hertz deal brought the production total for the model year to 2,378. Included was a special run at the end, of six convertibles, which were given to Shelby associates.

Walter Hane of Maitland, Florida, took the national SCCA B-Production title in a GT 350, marking the second straight year Corvette was shut out of an honor its drivers had taken each year from 1957 through 1964.

The production and sale of more than 600,000 cars per model year was more than the competition could ignore; and from the first success of the Mustang on, crash programs were underway among other domestic manufacturers to get an entry into the field.

Results of those programs would arrive during the 1967 model year. General Motors sent its Chevrolet Camaro and a little later the Pontiac Firebird; Chrysler restyled its Barracuda, adding hardtop and convertible models; and sister Mercury launched another contender, the Cougar.

1967

To greet its new neighbors, Mustang put on a new suit. Original features and concepts were retained, but enhanced with new sheetmetal interpretations. Also there were changes in the powertrain department with the first adaptation of the FE big-block for the production Mustang.

Exterior dimensions started their climb to bigger things. Length went from 181.6 to 183.6 inches, width from 68.2 to 70.9 and tread from 56.0 to 58.0 inches for V-8 models. Wheelbase remained at 108 inches. Heights were more or less the same and weights averaged an increase of ninety pounds for cars with similar engines.

Roof modifications ranged from a slight change around the rear window for the two-door hardtop to a whole new design for the Fastback 2+2. Fastback changes were executed by extending the roof slant angle back to the end of the trunk. A large, flat rear window and set of smaller louvers complemented the change. The basic theme would also be used the next model year on the intermediate Fairlanes.

Grilles were larger, extending to the bumper. The Mustang horse and its corral were back, mounted on bars again, with a crosshatch pattern backing it up. GT-optioned cars retained their four-inch fog lights. Side sculpturing was enhanced with deeper lines and a pair of die-cast fake scoops on each side. The rear area between the taillights was concaved. Three individual taillight lenses were on each side.

Inside, there was a new dashboard with the twin-cove and full-width themes combined. The driver faced two big dials, somewhat low on the dash, and provisions for three smaller ones, higher on the panel. Standard setup had the speedometer in the left large dial and a combination of oil pressure and ammeter gauges in the right. The top three were fuel gauge, an optional place for a clock and temperature gauge.

When the optional tachometer was ordered, the right large dial was used and its gauges were replaced by warning lights. The speedometer only went up to 120 mph this time, but Shelbys got 140-mph units. Air conditioning could now be built into the dash. The console got a redo and attached itself to the dash.

Mustang got some competition in the pony car sweepstakes from Mercury in the 1967 model year. It brought out a line of Cougar two-door hardtops. A mid-year addition to the 1967 Cougar lineup was the XR-7. While the outside looked fairly similar to the regular models, a plush interior awaited the passengers.

Somewhat revised was the GT Equipment Group, which now could be had with any V-8. When equipped with automatic, it became the GTA, just like the Fairlane. Included in the GT goodies were the fog lights, side stripes similar to earlier models, dual exhausts (on 271-hp 289 and 390 engines) with each side having dual pipes exit below the splash pan, the Special Handling Package, power disc front brakes and F70x14 wide-oval tires.

There were both interior and exterior decor groups. Inside were special door trim, seat material and other niceties. The Exterior Decor Group meant, among other things, a hood with far forward louvers and built-in turn signal indicators.

Engine-wise, the 1966 lineup was left intact (the 271-hp 289 was now called the Cobra V-8 and mid-year was restricted to GT cars only), with a 390-cube big-block, rated at 320 hp, placed at the top of the totem pole.

There was now room for a big-block and the feeling was it would be needed to combat the competition, which would be offering engines up to 400 cubic inches in the coming model year.

Mustang was the second Ford to get the 390 as a stop-gap performance measure against the competition. Fairlane also only had small-blocks through 1965, but received a 390 in the 1966 models. It didn't get the job done in the Fairlane and repeated its performance in the Mustang.

Ford first brought out the 390 for its 1961 big cars and it was built in one form or another ever since, but the last time it could be considered a performance engine was in the early 1962 models before it was replaced by the 406.

In Mustang trim, the 390 was very sedate. It carried a rating of 320 hp and while being reliable, it had breathing problems in the upper rpm range. It also added 250 unneeded pounds, much over the front wheels.

Some fairly decent street acceleration numbers could be obtained, such as *Hot Rod*'s 0-60 run of 7.1 seconds and quarter-mile jaunt of 15.31 seconds and 93.45 mph in a GTA 2+2; but the 390 Mustang was no match for the Chevy Camaro 350. To make matters worse, Chevrolet added the 396 big-block to the Camaro option list not long after the start of the model run.

"The 390 adds little to straight-on acceleration," noted *Car Life*, adding that the main purpose of the 390 was to run the power accessories and air conditioning.

While ad copy writers were extolling the virtues of the 390 Mustang, word of mouth was not following.

Hot Rod interviewed the performance manager at Tasca Ford, Dean Gregson, at the end of the 1967 model year. "We sold alot of 390 Mustangs last fall and into the winter, but by March they dropped off to practically nothing. That's when the snow melted off the asphalt. In fact, we found the car so non-competitive for the supercar field, in a sense we began to feel we were cheating the customer. He was paying for what he saw advertised in all the magazines as a fast car, but that's not what he was getting."

Hot Rod even started a write-in campaign to tell Ford that its cars for the street were too slow.

As in the previous two model years, there were three basic versions of the Mustang for 1967; the 2+2, hardtop (both with redesigned roofs and new sheetmetal below) and convertible. This 2+2 has the optional GT Equipment Group which includes the road lamps in the grille. It also is equipped with optional styled steel wheels.

Transmissions for the 390 included a heavy-duty three-speed, which was mandatory yet cost extra; the four-speed manual gearbox and new SelectShift Cruise-O-Matic, which could be shifted manually or automatically. All three-speed manual transmissions were now synchronized in all gears and the British four-speed was no longer optional for the six.

Front suspensions got a revision with 2.5-inch-longer lower control arms and lower upper A-frames. The changes helped make some room for the 390 and future large engines.

Fighting Mustang for a piece of the action was the Chevrolet Camaro, which would be Mustang's strongest competitor. It came in notchback hardtop and convertible versions, but there was no fastback. Styling was fairly conventional and a bit rounder than the Mustang. At first, a 295-hp 350 was the top Camaro option, but soon a 325-hp 396 joined it. With Chevrolet's liberal building program, it wasn't long before 375-hp 396's found their way out the door. While these ate 390's for a snack, Chevy further hit Ford where it hurt with a small engine, the 302-cube Z-28.

Chevrolet correctly eyed the Trans-Am series as the battle ground for the sporty compacts and, using the bore of the 327 and stroke of the 283, came up with an engine a lot closer to the 305 limit than Ford's 289.

Camaro came out with the rest of the 1967 models in the fall. A few months later, Pontiac's version came along—the Firebird. It shared Camaro's 108-inch wheelbase and basic body, but had its own drivetrain. With its smallest V-8 at 326 cubes, it was over the Trans-Am limit.

Mercury's Cougar shared basic body structure with the Mustang hardtop, but on a stretched 111-inch wheelbase. Length came to 190.3 inches and width was 71.2, all larger than Mustang. There were no convertibles or fastbacks; however, a mid-year XR-7 version came along.

Cougars came with the two-barrel 289 as standard, four-barrel, 225-hp 289 optional and 320-hp 390 as the hottest option. A Trans-Am program was announced along with a twin four-barrel version of the 289, rated at 341 hp. No special Trans-Am option, like the Z-28, was announced, however.

Plymouth now had convertible, notchback hardtop and fastback Barracudas, matching Mustang. A 383, rated at 280 hp was the biggest-cubed regular production option.

The race for sales leadership was no contest, as more than twice as many Mustangs were built as Camaros, but Mustang's total fell 22.3 percent to 472,121. The trend would continue for many years to come, Mustang would outsell Camaro, but continue to lose ground from the previous year.

While Mustang's sales curve started to point down, just the opposite was true for the Shelby versions. As the cars became more civilized (or dull, depending on how you look at it), more people bought them. Even though there wasn't a Hertz order during the 1967 run, it hit a record 3,225.

The first Shelby Mustangs put performance and handling ahead of styling. That priority changed with the 1967 models. Once again the 2+2 was used as a basis, but this time more changes were made to the body. Shelby got its own front panels, grille and scooped hood, all of fiberglass. The hood was held down by pins. Twin seven-inch driving lights were at either end of the grille cavity, although early mod-

Mustang's potential in stock car racing wouldn't be realized for a while, as few organizations' rules permitted smaller cars to race. NASCAR had a Grand Touring division in 1968 for the small pony cars. Donnie Allison (left) drove a 1967 Mustang for Jerry Huggins (right) and finished tenth in points that year and won five events, to account for Mustang's entire total on the circuit.

This dashboard, with two large dials and three small ones, was used in the 1967 and 1968 model years for Mustang. This convertible has several options including the console which was integrated into the dash, wood trim, radio, Cruise-O-Matic and convenience control panel in the center with four warning lights.

els had them paired at the center before they were determined to violate state laws in some cases.

Below the Mustang front bumper (which was two inches farther forward due to the design) was a fiberglass valance panel with a scoop below the grille for more air. Scoops were the theme on the rear quarter of the car as one covered the louvers with openings for air at either end. Another for brake cooling replaced the fake Mustang units on the rear fenders. A fiberglass deck lid and quarter-panel extensions combined to form a rear spoiler. The panel below the trunk lid/spoiler had a pair of long rectangular taillights which took up most of the area, except for the fuel filler.

Inside, a 140-mph speedometer and tachometer came standard while the Mustang's missing amp and oil-pressure gauges were mounted in the center, beneath the dash. A roll bar with inertia reel shoulder harnesses was standard, otherwise the interior was "fancy Mustang" with Shelby identification.

Besides the new styling, the biggest news for the 1967 Shelby was the addition of a second model, the GT-500. It came with Ford's 428, topped by a pair of Holley 600-cfm four-barrels on an aluminum intake manifold for a rating of 355 hp at 5400 rpm. Had a similar setup been available in the Mustang, no doubt the complaints would have been lessened.

Testing a GT-500 with a four-speed, *Motor Trend* found it capable of sixty from a start in 6.2 seconds and quarter-mile run of 15.42 seconds, trapping at 101.35 mph.

A few GT-500's were also equipped with a 427 Medium Riser, but the option was never advertised. The GT-350 continued to offer the 306-hp 289 and again a few got away with Paxton superchargers. GT-500's came with Ford's top loader four-speed or C-6 Cruise-O-Matic, while GT-350's got the same four-speed, or C-4 version of the automatic.

Fred Van Beuren of Mexico City, gave the GT-350 its third consecutive B-Production championship in the American Road Race of Champions finals at Daytona in November 1967, but the competitors were still using the 1965 and 1966 racing models, staying away from the heavier 1967 versions.

Mustang repeated as the Trans-Am champion, but not without a battle, as it only won four events out of twelve, all by Jerry Titus in a Shelby-prepared hardtop.

Strong competition came from the factory Cougar team, under the guidance of Bud Moore, and Roger Penske Camaros with some backing from Chevrolet and a strong lead driver, Mark Donohue.

In the season opener at Daytona, the Cougars were the fastest, followed by the Camaros and the Shelby-entered Mustang notchbacks. But when it was all over, Bob Tullius won in his 1966 Dodge Dart with a 273-cube V-8. Tullius let the faster leaders drop out.

Chrysler had withdrawn its support of Trans-Am racing for 1967 and despite the win, Tullius didn't get backing.

At Sebring, Florida, where handling is more important and horsepower less of a factor than at Daytona, Titus won his first race for Mustang. Donohue was second in a Camaro.

Cougar flexed its muscles with a pair of back-to-back wins at Green Valley and Lime Rock, Connecticut. Dan Gurney took the first and Peter Revson subbed for Parnelli Jones to take the second. Titus took his second win at Mid-Ohio, but Cougars still held the point lead.

Revson helped Cougar's total with a win at Bryar, New Hampshire. Donohue scored Camaro's first win at Marlboro, but Titus took consecutive wins at Colorado's Continental Divide Raceways and Crow's Landing at the Modesto Naval Air Station in California.

NASCAR driver David Pearson was called in to help Cougar and did so by taking the Mission Bell 250 at Riverside International Raceway in California. Team driver Ed Leslie finished second and gave Cougar the point lead.

Donohue gave a preview of things to come by winning the final two events at Stardust International Raceway in Las Vegas and at Pacific International Raceway in Kent, Washington. Ronnie Bucknum was second in both in a Shelby Mustang and wrapped up Ford's second straight title.

Points were awarded for the best nine finishes of the twelve-event season, and when it was all figured out, Ford drivers tallied sixty-four points, just two more than Cougar. Camaro was right behind with fifty-seven.

At the end of the season, Mercury withdrew its factory support from Moore, but that wouldn't be much consolation for the tough 1968 battle.

In drag racing, Mustangs were in for an off-year in 1967. NHRA changed its class structure and eliminated the factory experimental classes and expanded super stock.

There were some Mustang funny cars, but with only a 390, Mustangs weren't factors in S/S. Ford drag hopes were pinned on a batch of 427 Fairlanes.

One Mustang did attract some attention at the 1967 NHRA Nationals however. George Montgomery of Dayton, Ohio, gave up his AA Gas Willys body for a fiberglass 1967 Mustang and with supercharged sohc 427 power and a C-6 automatic, he easily took AA/GS honors with an 8.53-second run and trap speed of 166 mph! It was reported that lower center of gravity, longer wheelbase and reduced air resistance of the Mustang made it faster than the Willys.

1968

It was evolution time instead of revolution time when the 1968 Mustangs made their appearance. For 1968, mechanical changes, real and imagined, would take the spotlight, while styling was in a holdover pattern.

After a relatively simple early roster, the Mustang engine scheme of things began to get complicated starting with the 1968 models. Part of the reason was a set of new federal emission standards, and part was jockeying to improve racetrack and street performance.

Starting at the bottom, the 200-cube six got a compression cut from 9.2 to 8.8:1 and rating drop to 115 hp. Only one version of the 289 returned for 1968, a two-barrel setup rated at 195 hp.

New on the block was a 302-cube V-8 which had the same bore (four inches) and stroke (three inches) as the Chevy Z-28. It was achieved by a 0.13-inch stroke increase for the 289. For Mustang buyers, one version of the 302 was readily available, with 10:1 compression and four-barrel carb. Advertised horsepower listed at 230 at 4800. (We'll get back to other 302's a little later.)

At the start of the model year, one version of the 390 was listed, the four-barrel combo, listed at 325 hp, up five from the year before. A 280-hp two-barrel 390 joined it in November.

Car and Driver conducted a comparison of six 1968 pony cars. A 2+2 fastback with a 390 was chosen to represent Mustang. "The 390 engine has a reputation for being a bit of a stone," it said. "The Mustang displayed a lot of understeer in the handling tests, but with considerable applications of power in the tail could be brought around." *Car and Driver* concluded Mustang was resting on its laurels and rated it last. A Pontiac Firebird 400 HO won.

The pinnacle of the early model-year powerplant list was the engine with the highest advertised rating in Mustang production history, before or since: the Cobra 427 at 390 hp. It wasn't quite the fire-breathing V-8 of old that gave its all in drag, stock and sports car racing, as it had to cope with its first set of hydraulic lifters, a mild cam and 10.9:1 compression. To get by the tail-pipe sniffers, it only could be had with Cruise-O-Matic.

By December 1967 the 427 was deleted from the option list, never to return. However, its replacement turned out to be a worthy one—the 428 Cobra Jet. This same engine that Fairlane also got mid-year found its happiest home in the Mustang. Its low advertised horsepower rating of 335 at 5400 rpm got it slotted by the NHRA into C/Stock, where it thrived. Cobra Jets with bigger tires and other minor modifications qualified for SS/E.

Originally, when it was introduced for the 1966 models, the 428 was supposed to bring civilized characteristics to the seven-liter size, leaving the performance chores to the 427. However, in emission tests, longer stroke engines had an easier

Contents of the GT Equipment Group changed for the 1968 Mustangs. New items on this 2+2 include a C-stripe on the side, badge on the front fender and styled steel wheels with GT emblems on the hubcaps.

time, so the 428 was reevaluated as to what it could do to turn the wheels at rapid speed.

The 428 Cobra Jet turned out to be a compromise between the normal 428 and all-out 427 in components, output and price.

Taking it from the top, a Holley 735-cfm four-barrel sat above a cast-iron intake, similar in design to the aluminum Police Interceptor unit. Heads were like those on the 406 and had 2.06-inch intake valves and 1.625-inch exhaust, up from 2.022 and 1.551 for the normal 428. Compression was 10.6:1.

A few racing Cobra Jets were made for super stock competition with solid lifters instead of hydraulic ones, lighter valves and higher 11.4:1 compression with pop-up pistons.

And then there was Ford's Trans-Am engine for 1968. The 302 size was just right for the SCCA circuit and matched Chevy's Z-28 inch for inch. Ford planned to offer an option similar to the hot Chevy and qualify it for Trans-Am racing. It would have two four-barrel carbs and be similar to the Trans-Am 289's used in the 1967 Cougars. Compression was to be 11:1 and availability only in GT-optioned cars with four-speed transmissions.

Initially, the sales brochure said the engine would be on special order only and would have a rating of 345 hp. A reissue of the sales brochure dated December 1967 still talked about the engine, but now the rating dropped to 306 hp, the same as the Shelby 289.

Neither of the engines materialized, but then a third high-performance example was announced, the 302 Tunnel-Port. It used a principle borrowed from the 427 NASCAR engines that premiered in the 1967 Daytona 500. Rather than have rectangular ports limited in size by the space between the pushrods, it used large round ports for better flow and routed the pushrod tubes right through the middle of them. The result was better breathing. A new intake manifold was needed and the Tunnel-Port got an aluminum, two-plane intake job for a pair of four-barrels.

Actually the 302 and 427 Tunnel-Ports were different in basic design due to differences between the FE big-block and ninety-degree small-block. In the big-block, pushrods went through the intake manifold, so the pushrod tubes went through the passage in the intake rather than the head. In the 302 the tubes went through the passage in the head, as the intake ended before it reached the pushrods.

Another interesting feature of the 302 Tunnel-Port was the valvetrain, which utilized rocker shafts instead of studs for the rocker arms, as the regular 302's did. The shafts were said to be better suited to the higher speeds expected.

If there were suspicions that Ford underrated the Cobra Jet, they were nothing compared to the 302 Tunnel-Port, which got an advertised 240 hp at 5000 rpm and torque of 310 pounds-feet at 3000. That was only ten more horses than the regular four-barrel 302 and thirty-one fewer than the High Performance 289 that had less equipment to get out the horses.

Like the Cobra Jet, there were to be two versions of the 302 Tunnel-Port, one for the street with hydraulic lifters and 10.5:1 compression and a racing version with solid lifters, 12:1 domed pistons, stronger undersides, headers and transistor ignition. No rating was given for the racing engine.

Availability of the 302 Tunnel-Port was open to question in more ways than one. Both assembly-line and dealer installation were kicked around, but when it came down to it, neither won. Ford convinced SCCA that it would be an option for the Mustang and readily available and the engine was approved. Production plans

Cougar wasted little time getting into the performance field and this 1968 GTE with a 427 V-8, raised hood and styled steel wheels clearly demonstrated that.

fell through with Ford claiming the engines weren't ready for the public yet, leaving the Trans-Am cars with one of the most unusual of all Ford small-blocks.

Ford was still having credibility problems between its extensive racing and high-performance publicity programs and what it was actually selling for the street—and the Tunnel-Port fiasco didn't help one bit.

Compared to the engine situation, the cars themselves were rather calm. Only minor styling changes were made in the 1968 models, but there was an interesting mid-year addition, the Cobra Jet option.

Model selection and body panels remained basically the same. Grille changes included omitting the cross bars again, replacing them with a perimeter chrome design. Side dummy scoops changed from twin units to long single pieces and the government-mandated side-marker lights (front) and reflectors (rear) were mounted. The Exterior Decor Group fell off the option list, but the vented hood remained an extra.

Inside, the new models used the same basic dash and interior layout with new trim and a new steering wheel.

Several revisions were made in the GT Equipment Group for 1968. The four-inch fog lights were at the outer extremes of the mesh grille. Side trim changed with a tapered C-stripe following the body lines, chrome rocker panel trim and new GT badges. The old stripes were also available.

In the rear was a pop-open gas cap with GT lettering. Dual exhausts still had twin pipe outlets below the splash panel. Styled steel wheels with GT emblem hubcaps set off the package handsomely. They were six inches wide and held F70x14 white-stripe, wide-oval tires.

GT availability included a selection of engines 302 cubic inches and up.

Early in the model year the Sports Trim Group was also an option package. It contained wide flat-black strips on the required louvered hood, wood-grain instrument panel trim, wheel lip moldings, special vinyl seats (except on convertible) and styled steel wheels and wide-oval tires, the latter two on V-8-engined cars. By December the package was deleted from the option list, but the items were all available as individual options.

Some mid-year option packages livened up the looks of the Mustang, but did nothing for performance. The Sprint came out in the spring with a GT-like stripe, pop-open gas cap and full wheel covers. In addition, V-8 models got GT fog lamps, styled steel wheels and wide-oval tires.

More impressive was a regional model, the California Special which was sold through Ford dealers in the southern part of the state. Normally, regional models have a little extra trim, a badge or two and hold little interest, but the California Special was quite the exception. Available only in hardtop form, it wasn't tied to any engine option—you could get six or V-8. It was more or less a cross between a Mustang and a Shelby as far as visual items went.

Starting from the back for a change, it used a fiberglass trunk and spoiler from a 1968 Shelby convertible, along with Shelby taillights. Nonfunctional side scoops, also of fiberglass, replaced the conventional Mustang units. Grille ornamentation was gone with a blacked-out mesh background. Aftermarket rectangular fog lamps were mounted at either end, like the Shelby.

The louvered hood also had twist-type hold downs in the front corners. Striping was generous with side stripes ending at the scoop with the lettering GT/CS reversed out. The rear of the spoiler and quarter panel extensions also got a set of stripes.

Denver-area Ford dealers also got their version of the car called the High Country Special. A decal replaced the GT/CS lettering on the scoop.

All this was fine for those who delighted in things visual, but for the performance-minded, Ford brought out the first of several special performance Mustangs, the Cobra Jet.

To let the world know about its new Cobra Jet performance engine the Cobra Jet option package was presented in spring. It utilized a fastback with the GT option. Naturally it came with the big 428 and heavy-duty underpinnings.

The unique feature of the Cobra Jet was a functional scoop on the hood, located within a wide stripe that went from the leading edge of the hood to the windshield. Underneath it was an air cleaner that opened up a panel under the scoop when full throttle activities were carried out. Four-speed cars got an 8000-rpm tachometer, while automatic buyers had to pay extra for it.

Hot Rod's Eric Dahlquist had great fun in 1967 putting the needle to Ford for such slow cars on the street. Ford turned the tables on him by getting him to test a preproduction Cobra Jet. It was one of the handful of racing-engine Cobra Jets, which were far friskier than the street variety and had just been set up for the NHRA Winternationals by Holman & Moody/Stroppe on the West Coast.

There was no sound deadener, it weighed 3,240 pounds and blew Dahlquist into the weeds. A 0-60 sprint of 5.9 seconds and quarter-mile cruise of 13.56 seconds and 106.64 mph had him stumbling all over himself to compliment Ford. "Quite frankly, it is probably the fastest regular production sedan ever built," Dahlquist noted. This opportunity was not lost on Ford, as ads were taken out in the buff books quoting its one-time critic.

Slotted in SS/E and SS/EA, Cobra Jets did their thing at the Winternationals in Pomona in February. Al Joniec used his class handicap to take super stock eliminator honors, after winning his class over runner-up Hubert Platt, also in a Cobra Jet.

In the automatic competition, factory driver Dick Landy put his Dodge ahead of runner-up Gas Ronda, aboard another hot CJ. Platt was considered a likely contender for C/SA honors in his Cobra Jet, but he didn't hear the call for his class run-off due to confusion in the pits and Dave Kempton made an uncontested run for the title in his early Plymouth 383.

Cobra Jets did well during the 1968 drag racing season. Jerry Harvey capped it with class honors for SS/E in the NHRA Nationals at Indianapolis Raceway Park.

As if there wasn't enough competition in the sporty compact field (now being called "ponycars" after you know what), American Motors joined the fray with its Javelin. Mid-year a shortened version with just two seats came out, the AMX.

American Motors took a familiar path to promote the Javelin, Trans-Am racing. With Mercury Cougar gone and Chrysler still inactive, the players in the Trans-Am series for 1968 would be the Shelby-entered notchback Mustangs, Camaros from Roger Penske and the new Javelins with the team headed by former sports car racer Jim Jeffords.

Camaros got a new cross-ram twin four-barrel intake and would go wheel-to-wheel with the Mustang Tunnel-Port 302. While the Camaro would do just fine, the Tunnel-Port turned out to be troublesome.

As a result, Mustang would not get a third straight manufacturers championship. Camaro, on the strength of Mark Donohue, won ten of the thirteen races on the schedule and easily took the title with 105 points to sixty-three for Mustang, which accounted for the other three flags. Javelin's rookie season was good for lots of learning, and third place at fifty-one points.

In the season finale at Kent, Washington, where Donohue scored the final win for Camaro, there was some driver switching going on. Jerry Titus left Shelby to campaign a Chevy-powered Firebird with teammate Craig Fischer. Peter Revson left the Javelin team to fill in at Ford and Lothar Motschenbacher moved into the vacated Javelin driving seat. It was all part of the jockeying for final points, as there was a battle between Mustang and Javelin for second.

Mustang may have fallen from its place at the top of the Trans-Am standings, but it was still ahead in sales in the class of cars it created.

In the 1968 model year, Mustang production totaled 317,404 compared to 235,151 for runner-up Camaro. However, Mustang's production took its biggest tumble yet, falling 32.8 percent from 472,121 the year before, helped by a long strike. The 1968 model year was not all that bad overall, as production rose 9.6 percent for domestic makes.

Once again the trend was bucked by the Shelby Mustang, which had a thirty-eight-percent increase in orders to 4,450. However, as had been happening in the past, Shelbys were more and more Mustang and less Shelby.

A mid-year addition to the Mercury Cougar line for 1968 was the XR7-G, with the letter G referring to Dan Gurney, who signed up to promote Lincoln-Mercury cars for 1968. Among the features were a fiberglass hood scoop, road lamps, hood pins, styled wheels, quad exhaust pipes and special ornamentation.

Carroll Shelby stands between a pair of cars which carried his name, but not necessarily his heart, the 1968 Shelby Cobras. The 1968 models were the first full-year run that were not built in Shelby's factory in California. For the first time, a Shelby was mass-produced in convertible form in the 1968 model year. The GT-500 soft top sported a padded roll bar and the 1968 styling which was becoming more important than the performance.

An indication as to how far the dilution had progressed was the announced move of production from Carroll Shelby's shops at the Los Angeles International Airport to the A. O. Smith factory in Ionia, Michigan. While press releases proclaimed that "We at Ford are looking forward to an even closer relationship with Carroll," it marked the beginning of a split between the two. Shelby Automotive was moved to Detroit, while Shelby Parts and Shelby Racing remained in Torrance, California.

Starting with the 1968 models the cars were called Shelby Cobra. Series designations remained the same at GT-350 and GT-500, but there were powertrain changes, the addition of convertible models and a mid-year GT-500KR addition.

Convertibles were available in both the GT-350 and GT-500 series and came with a padded roll bar. Other than a few specially built models, it was the first production soft top for Shelby.

Like Mustang, the 1968 Shelby Cobra was in the second year of its styling cycle and only had detail changes from the 1967's.

Hood design was changed, moving the scoop from the center of the hood to a twin setup at the front. Round fog lights were replaced by rectangular units of the aftermarket type. Lucas and Marchal lights were used. Taillights had detail changes with a chrome-like covering that sectioned them into six segments.

Power for the GT-350 changed from the 306-hp 289 to a milder 250-hp 302. At first cars came with a cast-iron intake and Autolite four-barrel, but a running change switched over to an aluminum intake and Holley 600-cfm carb. The setup was retrofitted on some earlier models.

In the GT-500 department there were three different powerplants. Early GT-500's came with the 428, rated at 360 hp and very similar to Ford's Police Interceptor 428. They had an aluminum intake and 715-cfm Holley four-barrel.

A limited run of GT-500's came with 427's aboard with medium-rise intakes. They were rated at 400 hp at 5600 rpm and had hydraulic lifters and only came with automatic transmissions.

Late in the model run the GT-500KR replaced the GT-500. It came in both fastback and convertible models. The "KR" stood for King of the Road, which was to be the result after the installation of the Cobra Jet 428 in place of the 360-hp 428.

The KR carried the same advertised rating as the Cobra Jet in the regular Mustangs. It got new emblems for the fenders and dash and recognition with revised lettering in the side stripes. Another feature was the channeling of air from the front scoops to the air cleaner.

1969

Mustang's two-year styling cycle dictated major changes for the 1969 models and that's exactly what happened. However, not only were the bodies redone, but there were also engine changes with a couple of mid-year models that were to be the high-water mark for performance in Mustang's history.

Anyone seeing the 1969 Mustang for the first time would have little trouble identifying it. Familiar themes were carried out and exterior dimensions increased again. Wheelbase remained the same at 108 inches, but length grew from 183.6 to 187.4 inches. Width progressed from 70.9 to 71.3 inches.

Three body styles continued with new roofs for the fastback (now called SportsRoof) and notchback two-door hardtop. Windshields were sloped more on all models. Front vent windows disappeared.

At the start of the model year, the hottest Mustang was its new performance model, the Mach 1. Utilizing the SportsRoof body, it came with a standard 250-hp 351. This one has the optional 428 Cobra Jet with shaker hood scoop. Stripes, styled steel wheels and hood pins were all part of the standard equipment.

The SportsRoof lost its vent system and gained swing-out rear quarter windows. The roof slanted back, again contained a large rear window and ended just before the new spoiler kick-up on the trailing edge of the trunk. A phony scoop was located high on the rear quarter panels, just below the rear side window. Hardtops were smoothed-out with the formal roof lines sweeping back more.

The familiar central grille and flanking headlight coves were retained, but a second set of headlights was installed at either end of the plastic grille, giving Mustang its first (and for quite a while, last) set of quad headlights. The valance or splash panel was changed to form a large scoop beneath the front bumper with parking/directional lights at either end.

Hoods were redesigned with a long, narrow power dome running down the center.

Side styling was simple with a line extending from the front to just into the rear quarter panel. On SportsRoofs it ended in the high scoop, while on hardtops and convertibles, a bolt-on nonfunctional scoop extended below the line, similar to that on earlier Mustangs.

Rear styling retained the cove below the trunk with three vertical taillight rectangles at either end and the gas filler cap in the middle.

Insides were changed with recessed panels for driver and passenger, surrounded by heavy padding. For the driver, four round dials, two larger ones in the center with two smaller on the ends, contained the necessary gauges. In standard layout, (from left) came the ammeter, 120-mph speedometer, combination fuel and temperature gauges and oil pressure gauge. When the optional tachometer went into the right larger pod, the temperature gauge went to the left unit and fuel gauge to the right. Warning lights for oil pressure and ammeter were in the tachometer pod.

Consoles were redesigned and no longer blended into the instrument panel. The rear portion formed an armrest between the front bucket seats. When the cars got clocks, they were installed on a panel in front of the passenger, above the glove compartment.

A pair of new models, each aimed at a different market segment, joined the traditional three base models. Performance fans got the Mach 1, a SportsRoof with special equipment. A 250-hp 351 was standard and options went up in horses and cubes from there.

External features of the Mach 1 included stripes along the side of the car, extending from behind the front wheel to the rear side-marker light and another on the rear spoiler with the name reversed out. Hoods received a large, flat-black panel that extended back to the cowl. A fake scoop was standard and a real one, which we'll get to shortly, was optional. Hood pins were standard for the front corners, but could be deleted at the buyer's request.

Ford's neat color-keyed racing mirrors were standard, as were chromed styled steel wheels and plain hubcaps. There also was a pop-open gas cap and a narrow rocker panel molding.

Inside were high-back bucket seats, a console, wood-grain covering on the instrument and passenger panels, Mach 1 emblem on the passenger panel, which contained the standard clock and special door trim. A handling package with stiffer springs and shocks was standard and extra insulation was added.

A redesigned instrument panel was used on the 1969 Mustangs with twin coves. This is the Grande interior—a new luxury model—with console, wood trim and clock in front of the passenger.

As the Mustang grew, so did the Cougar. This 1969 CJ 428-optioned hardtop was 3.5 inches longer and nearly three inches wider. Mercury's pony car paralleled Mustang in performance offerings. Note the large functional hood scoop, hood pins and mag-style wheels.

Aimed at the luxury-oriented buyer was the Grande, which came in two-door notchback form. It featured a fancy interior, complete with cloth centered bucket seats and Grande nameplates on the rear roof pillars.

Back for its final appearance in the Mustang lineup was the GT Equipment Group, which was overshadowed by the Mach 1. It was available on all three base models and needed 351, 390 or 428 power. Stripes above the rocker panel were there again, but no longer had GT emblems. Styled steel wheels had GT emblems on the hubcaps. Cars with dual exhaust got the twin pipe outlets. The pop-open gas filler cap also got GT emblems. GT hoods received the fake scoop standard, which like the Mach 1 had directional-signal indicators built into the back, and hood pins. Heavy-duty suspension completed the package.

At the start of the model year, engine availability was the most complicated yet for Mustang and it would get even worse before the end of the model year.

Standard for all models except the Mach 1 came the 115-hp 200-cube six. New was an optional 250-cube six, rated at 155 hp.

First V-8 in the arsenal was the 302, now reduced to two-barrel, 220-hp form after spending the previous model year with a four-barrel and 230-hp rating.

Standard on the Mach 1 and optional on other models was the 351 Windsor, with a two-barrel carb, 9.5:1 regular-gas compression ratio and rating of 250 hp. Optional for all models was the four-barrel, 10.7:1 version, at 290 hp.

Returning for its swan song engagement in a Mustang was the 390, in one version, with a four-barrel, 10.5:1 compression and 320-hp rating.

Most potent, at least at the start of the model run, was the 428 Cobra Jet with its 335-hp rating. There were two versions, with and without Ram-Air. Both carried the same advertised output. Ram-Air meant a functional hood scoop, which Ford called the "shaker." It was mounted right to the air cleaner and protruded through a hole in the hood. When the engine gyrated, so did the scoop.

Transmission availability included an all-synchro three-speed standard up through the 351 four-barrel, four-speed wide-ratio optional for the 302 through 390 and four-speed close-ratio optional for the 351's on up. SelectShift Cruise-O-Matic was optional for all engines. Differential ratios ranged from 2.33 to 4.30:1, depending on application.

Reaction to the Mach 1 was generally good in the motoring press. *Car Life* called it "The first great Mustang," later naming it the best pony car of 1969. They tested one equipped with a Cobra Jet, shaker, automatic and 3.5:1 rear axle. It blasted the quarter-mile in 13.9 seconds, hit the traps at 103.32 mph and wound up to sixty in 5.5 seconds. "The Mach 1 is the quickest standard passenger car through the quarter mile ever tested," the report noted, saying that it beat the previous mark held by a Hemi-equipped Plymouth.

Ford's answer to the Chevrolet Camaro Z-28 was the mid-year Boss 302 version of the Mustang SportsRoof. It came standard with a tough 302 V-8, rated at 290 hp. The black hood panel, side stripes, front spoiler and flat-black headlight and rear deck trim were standard. Options on this copy include rear deck spoiler, Sports Slats over the rear window and chrome Magnum 500 wheels.

Car and Driver put the Mach 1 on the cover of its November 1968 issue, liked its styling, was impressed with its acceleration with 3.91 gearing, but blasted the handling with the big 428 up front. "The beak-heavy machine just won't corner with any dignity at all. Does it understeer? Yes sir, yes sir, three bags full."

As it turned out, the Mach 1 served to warm up the audience for the main attractions, the Boss 429 and Boss 302. Both were built to legalize Ford's role in racing during the 1969 season, though each had a specific, nonrelated purpose.

First out of the chute was the Boss 429, which was the most unusual of the pair. Its purpose was to get the big 429 engine legalized for competition in NASCAR Torinos and Cyclones. Ford had to build at least 500 of them in a passenger car to qualify. Instead of installing them in intermediate Torinos, the Mustang was chosen.

First, let's look at the engine called Boss 429, or Cobra Jet HO at different times. Ford introduced a new thin-wall big-block, called the 385 series, in the 1968 Thunderbird. It had a bore of 4.36 inches and stroke of 3.59. The new casting had wider bore centers than the old FE series big-blocks. Rating was a mild 360 hp at 4600 rpm. Low-stressed leisurely performance was the early 429 goal.

However, long-range plans called for the 429 to replace the FE engines (427 and 428) for high-performance and racing applications.

In NASCAR competition Ford's wedge 427 had long battled Chrysler's 426 Hemi and not always on equal terms. There were several updatings to the Ford engines over the years, but they still lacked the advantage of the deep-breathing hemispherical heads of the MoPars.

In Boss 429 form, Ford sought to have a big-inch, ultra-powerful engine that could take care of the Hemis and also, in drag racing applications, replace the complicated sohc 427.

While the FE engines were of Y-block construction, the new 429 eliminated the skirt below the centerline of the main bearings. Boss blocks got four-bolt main bearing caps on the first through fourth mains; but both sets of bolts went down into the web, instead of being cross-bolted from the outside of the block, like on the 427's. The fifth main only got two bolts.

One unique feature of the Boss was the aluminum heads, which were dry-decked with O-rings used around the combustion chambers and no gaskets. Large, oval intake ports led to large valves in hemi-type combustion chambers. Valves were mounted on individual rocker shafts bolted to the head. Another unique Boss 429 feature was the steel billet crankshafts, which were cross-drilled and had grooved main journals.

There was more than one version of the Boss 429 and even more than one version of the street engine, although there was no change in the ratings. Since Ford couldn't call its new creation a hemi, other names were used. Among them Shotgun 429, Blue Crescent 429 and Boss 429. Outside the factory it also got the tag "twisted hemi."

The first 279 cars built had the S version of the engine. It had a hydraulic lifter cam with stronger rods. The later T version used different rods and pistons and came with both hydraulic and mechanical lifters.

The ultimate performance Mustang came out as a mid-year model in 1969, the Boss 429. The specially constructed car housed the NASCAR Boss 429 engine. Compared to the Boss 302, the exterior of the big Boss was plain. Seen here are the functional front hood scoop, front spoiler, chrome Magnum 500 wheels and simple Boss 429 decal behind the front wheel.

Racing engines featured stronger undersides, domed pistons and larger valves. A single four-barrel 735-cfm Holley fed the horses through an aluminum intake manifold. Advertised output was 375 hp at 5200, while torque came to 450 pounds-feet at 3400. Again, Ford was playing conservative with its output figures.

Completing the driveline was a high-performance clutch, Ford top loader close-ratio four-speed, Traction-Lok differential with 3.91:1 gearing standard and optional ratios ranging from 3.5 to 4.30:1.

Serving as a basis for the Boss 429 was the SportsRoof, which remained remarkably free of frivolous exterior decoration. Most noticeable was the large scoop in the center of the hood. Instead of being controlled by the throttle opening, a cable to the passenger compartment gave control of scoop opening to the driver.

There was a spoiler below the front splash pan, front fenders were flared for clearance for the big tires. There were no trick stripes, only simple Boss 429 decals on the front fenders behind the front wheel openings. Chrome Magnum 500 seven-inch-wide wheels carried huge F60x15 tires.

In order for the big Boss engine to fit, the front suspension was modified, with spring towers being moved an inch farther out for clearance. A-arms were lowered an inch, dropping the front of the car and giving it a rake. Suspension was heavy-duty with a rear stabilizer bar added.

Other details included a trunk-mounted battery, engine oil cooler, power brakes with 11.3-inch front discs and power steering. Some items were listed as mandatory options. Inside, the Boss 429's had a standard tachometer, high-back bucket seats and console. Interiors were black.

Due to the special requirements of modifying the front suspension, the cars could not economically be built at a regular Ford factory. Instead the job was farmed-out to Kar Kraft in Brighton, Michigan, which worked on Ford's special performance projects. A small assembly line was set up and in January of 1969, production of Boss 429's began. A total of 858 models for 1969 were built.

With mandatory options, the Boss 429 listed for $4,798, which was somewhat expensive for 1969, but reportedly well below what it cost Ford to make it. Boss 429 engines were also installed in a couple of Mercury Cougars, but it was not a regular option for Cougar. The job of the Boss 429 was to handle the NASCAR obligations and take in a little drag and street racing on the side. Next up was the Boss 302, which was aimed at homologating all the goodies for Trans-Am racing.

Despite its experience, Ford had a terrible time in Trans-Am in 1968. Its Tunnel-Port 302 didn't work as well as planned on the track and after much hoopla, never got to the street. This time around, Ford was determined to do things right.

Once again the SportsRoof served as the basic car and like the Boss 429, a special engine was the focal point.

As indicated by the name, the Boss 302 utilized Ford's 302-ci V-8. Modifications were considerable with four-bolt caps on the second, third and fourth main bearings and two-bolts on the other two, solid lifters, a 290-degree cam and heads with canted valves to allow large ports. There also was a windage tray over the oil pan, 10.6:1 compression and a 780-cfm Holley four-barrel on top of a dual-plane aluminum high-rise intake. Intake valves were a large 2.23 inches, while exhausts were 1.71.

A mid-year addition to the Mercury Cougar line for 1969 was the Eliminator. It was strong on visual effects with large hood scoop, front and rear spoilers, stripes and styled steel wheels. Standard engine was a four-barrel 351. One of the options was the Boss 302 V-8.

Actually, the engine breathed a little too well for the street and there were problems with bottom-end output and fuel mileage.

Advertised output, while not as ridiculous as the 240-hp rating for the Tunnel-Port, was still conservative at 290 hp at 5800 rpm. Torque was the same number, 290 pounds-feet at a high-revving 4300 rpm.

To keep owners out of trouble, Boss 302's were equipped with a rev limiter which was attached to the car's coil wiring. At 5800 the current started to be cut and by 6150 it was limited so the car couldn't go above that. It was felt to be needed due to the quick-winding characteristics of the engine. However, many owners felt they didn't need it and it was easily disconnected.

Moving on down the driveline were a 10.4-inch high-performance clutch, four-speed transmission with wide (2.78, 1.93, 1.36, 1:1) or close (2.32, 1.69, 1.29, 1:1) ratios, a three-inch driveshaft and nine-inch ring-gear-equipped differential. The standard ratio was 3.5:1, with optional Traction-Lok 3.5, 3.91 and 4.3:1 ratios.

Wheels and tires were like those used on the Boss 429 except the wheels were argent. Front suspension was in stock position and the rear suspension lacked the big Boss anti-roll bar, but suspension was strictly heavy-duty. Steering was an ultra-quick 16:1 with manual standard and much-needed power optional.

Production of the Boss 302 started later than the 429 with availability starting in early spring. Without special front end chassis work, the 302 could be made on a regular production line. Its exterior trim was more gaudy and aimed at what Ford thought was the typical high-performance buyer.

Rear quarter panel scoops were filled in. C-stripes marked the sides starting from the rear bumper and going forward to the front wheel opening where they widened, shot up with Boss 302 lettering in the middle, then jutted back in a narrow tapered fashion, ending just behind the door. Front fenders were also changed to allow for the big tires.

Hood, cowl and headlight housings got flat-black trim. A front spoiler looked like the one on the 429, but was larger. At the back there was flat-black paint on the deck lid, which blended well when the optional bolt-on rear spoiler and Sports Slats were added. There was no special interior for the Boss 302 as all regular Mustang options were available.

Mercury offered a counterpart to the Boss 302, the Cougar Eliminator (a 351 was standard with the hot 302 optional). Actually, it was a bit of a misnomer as the name pertained to drag racing, but when the small 302 went in, the heavier Cougar body made it anything but a threat in drag racing. It was more adapted to Trans-Am racing, but Mercury no longer had a team in that sport.

When they got to the street, the Bosses weren't all that boss. *Car Life* noted the Boss 429 looked good on paper, but was actually slower than the Mach 1 428 with a 0-60 run of 7.1 seconds and quarter-mile time of 14.09 seconds and trap speed of 102.85 mph.

A show model of the Boss 429 engine was loaded with chrome accessories, but did demonstrate the massiveness of the powerplant. In order to fit it into a Mustang, modifications to the front suspension were needed.

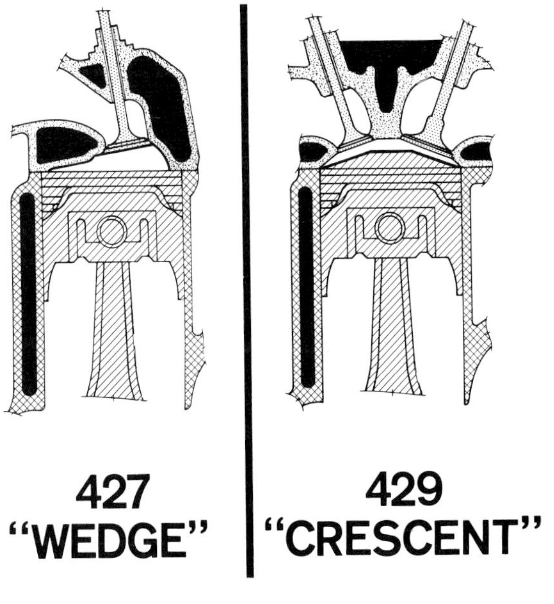

Ford couldn't call its new 1969 Boss 429 a Hemi, as Chrysler used that name, so it described its combustion chambers as "crescent" shaped, which from the side was correct. The difference between the Boss and the old 427 wedge engine is shown.

Of the two, the Boss 302 was the better-balanced, although it lacked cubic inches. *Sports Car Graphic* put it on the line by comparing the Boss 302 to its direct competition, the Camaro Z-28. Both had four-speeds, 302-cube engines and ratings of 290 hp. The Boss tested came with the wide-ratio gearbox and 3.5:1 gears compared to the Z-28's close-ratio four-speed and 3.73 gearing.

The magazine testers may have exaggerated a bit, but they published the same performance figures for both cars recording 0-60 in 7.1 seconds, 100 mph in 16.8 and the quarter in fifteen seconds at 96 mph. They noticed the Boss handled more competently. "The Boss may corner just a little better. Seat-of-the-pants feel and the cornering photos tend to indicate that it rolls a tad less and doesn't seem to ride as hard, which indicates that the Boss has softer springs or shocks and a harder anti-roll bar. The Boss also seems to understeer a little less." They also admitted that the Boss was a tad faster, but in a close vote chose the Z-28 as their favorite, mainly on the basis of its reputation. They closed with, "But don't get smug, Chevy! For a 50¢ difference we'll take the Boss—to win."

Car and Driver wrung out an engineering prototype and praised its handling and smoothness compared to the Z-28. It noted the weight distribution was 55.7 percent on the front wheels compared to 59.3 for the Mach 1 428.

"Without a doubt the Boss 302 is the best handling car to come out of Dearborn and just may set a new standard by which everything from Detroit must be judged."

"If the production cars are anywhere near as good as the prototype, the Boss 302 is easily the best Mustang yet, and that includes all of the Shelbys and Mach 1's," *Car and Driver* said.

Ford wasn't the only manufacturer playing with its pony cars. Plymouth was sticking 440's and Hemis into a few of its Barracudas. Chevrolet ran off a small batch of ZL-1 Camaros with aluminum 427's aboard. And Pontiac introduced a Trans-Am version of the Firebird and announced a 303-ci V-8 which met a similar fate as Ford's Tunnel-Port 302 trying to get to the marketplace.

Despite all the hoopla, sales of the 1969 Mustang continued the annual drop, but it was not severe, falling 5.5 percent to 299,824. Notable was the Mach 1's performance of 72,548 orders and the overall SportsRoof production of 134,438 for a whopping 44.8 percent of the total production. The production was the highest ever, before or since, for the fastback Mustang body style.

Camaro got closer to Mustang than ever before, with 243,095 being produced, but needed an extended model year to do it. Instead of ending in summer with the rest of the 1969 models, 1969 Camaros were produced late in the calendar year, as introduction of the 1970 models was delayed until after the start of calendar 1970. Some late 1969 models were titled as 1970's, but were counted in the 1969 tally.

Performance was the name of the game in 1969, and Ford played hard in most areas where the Mustang could compete.

To get things rolling, Ford President Bunkie Knudsen sent his buddy Mickey Thompson to the Bonneville Salt Flats to go after some records. Three modified

Powering the Boss 302 was an engine of the same name. It was toughened from top to bottom for the chore of Trans-Am racing. Its horsepower rating was a conservative 290.

Taking up the front row of the Laguna Seca Trans-Am in 1969 were Parnelli Jones (#15) and Bud Moore Engineering teammate George Follmer. The cars now carried Boss 302 striping. They did not carry the checkered flag, however, as Mark Donohue dominated the race and the series that season.

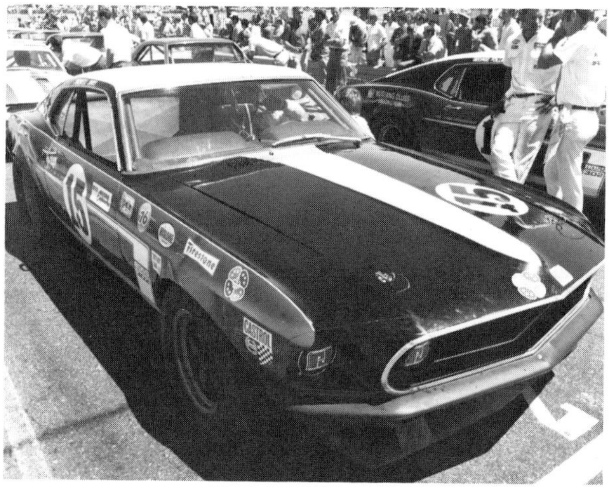

1969 Mustangs were used with power ranging from 302 to nonproduction 427 cubic-inchers. Thompson, drag racer Danny Ongais, writer Ray Brock and Bob Ottum drove and came back with 295 records, as certified by USAC. Highlighting the show was a twenty-four-hour run in a 302-powered SportsRoof averaging 157.663 mph over a ten-mile course.

Mustang's big show was the Trans-Am. The Boss 302 made homologation and for the first time, Ford had a fastback to race in the series. Ford had to build 1,000 Boss 302's to make it and with a reported 1,628 being made, it was home free.

As usually was the case, there were differences between the street and racing engines. While there were many minor ones, the major changes included dry-decking the heads, special pistons and rods and four-bolt mains all the way across.

After a dismal 1968 season, Ford was taking no chances and had two Trans-Am teams. Carroll Shelby had a pair of cars at the start of the season, wheeled by Peter Revson and Horst Kwech.

Bud Moore returned to the Ford fold after fielding Cougars in 1967. (He ran them in 1968 in NASCAR's new Grand Touring division and Tiny Lund won the title in one of them.) Moore had the strongest drivers, Parnelli Jones and George Follmer. Shelby's cars were blue with white trim, Moore's were red, black and white. Naturally all were Boss 302's. Both teams' cars carried identification for the National Council of Mustang Clubs, which Ford organized and backed.

The season opened May 11 at Michigan International Speedway with Parnelli Jones taking a win for Mustang in the Bud Moore car. Several top-name drivers were at Indianapolis on May 30, so substitutes were used at Lime Rock, Connecticut. Sam Posey drove Peter Revson's Shelby Team Mustang to victory, with Swede Savage bringing in Parnelli's car for second.

It appeared that with the Boss 302's running, the battle would be between the Moore and Shelby teams rather than between Camaro and Mustang. Ronnie Bucknum interrupted that in a Penske Camaro at Mid-Ohio on June 8, but Mustang bounced back with two more wins at Bridgehampton, New York, June 22 under the guidance of Follmer and at Donnybrooke in Minnesota on July 6, thanks to Jones.

Ford held a good lead in the standings at that point, forty to thirty-one over Camaro, but for Mustang fans the season might as well have ended there. Camaros won all seven remaining races with Donohue taking six and Bucknum one.

Mustang's downfall came August 3 at the St. Jovite event in Canada. All four team cars crashed and no points were earned. From there on, it was catch-up racing and the Mustangs never caught up. An early-season Camaro suspension problem was solved and, even though the Fords could still out-qualify the Chevys, the end result was a Penske Camaro in victory circle.

Camaro ended with seventy-eight points (based on the best nine finishes), with Mustang second at sixty-four. Firebird, with engine legalization problems, was a distant third with thirty-two. However, nobody was as bad off as Javelin, which couldn't do anything right during the racing season and only got thirteen points.

After the racing season American Motors did something right, it signed Penske and Donohue away from Chevrolet to operate the Javelin team.

After several years' absence, Ford organized a rally team, only this time the efforts were concentrated in the U.S. The effort paid off and Mustang drivers Rodger and Cathy Bohl brought Ford the SCCA Manufacturer Rally Championship for 1969, the first time Ford ever won the honor.

A side benefit of the Boss 429 was to be success in drag racing. A few cars were built especially for that purpose and NHRA slotted them in the top super stock class, SS/A, against the ZL-1 Camaro. Neither did all that well and they were downgraded to SS/C, where the Boss still had problems.

The hottest super stocks were the Hemi-powered Darts and Barracudas, but Mustangs and other Fords did fairly well in the lower classes. Mustangs were campaigned by the Ford Drag Teams that toured the country and visited dealerships conducting performance clinics.

With Mustangs going around flexing their muscles, their cousin, the Shelby, was doing anything but. Using the new Mustang body shell, the 1969 Shelbys had come full circle from their original intent. The early 1965 models *looked like* Mustangs, but had mechanical changes to make them competitive as sports cars. The new 1969 Shelbys had changes so they looked *different from* Mustangs, but mechanically they were just about identical to them.

After having the Cobra name one year, 1969 Shelbys lost it, as Ford assigned it to a budget supercar in the intermediate Fairlane lineup.

Shelby's claim to fame in 1969 was a fiberglass front end which divorced itself from the Mustang theme. All body panels forward of the firewall were fiberglass.

Grilles were full width with single headlights at the ends. A screen-like pattern filled the center with bright metal surrounding. A thin lower bumper was matched by chrome trim around the rest of the opening. A widely scooped valance panel with directional signals was under the grille. Beneath the bumper were the rectangular driving lights.

The hood must have broken some sort of record for scoops, having five inlets and/or outlets. The center one fed fresh air to the air cleaner while there was a pair of raised tunnels to either side which had scoops both at the front to let in cool air and at the back to let out hot air. Round hood fasteners were at the front corners. The redesigned front fenders also got scoops which were for the front brakes. Wide stripes continued down the side of the car at the halfway point.

More scoops were hung on the rear quarter panel. On the Shelby SportsRoof one was located high, as on the Mustang SportsRoof. It was in its traditional place on the convertible. Both were kind of functional, leading to the rear tire, but not directly to the brakes. Rear styling featured a more pronounced spoiler compared to the regular Mustang. A black inset was used to mount Shelby lettering.

Retained for 1969 were the 1968 taillights. In place of the license plate under the rear bumper was a pair of rectangular exhaust stacks. The license plate was moved up to cover the fuel filler. Special Shelby wheels were five-spoke mag types with aluminum centers and steel rims. They were seven-inches wide and held standard E70x15 tires. There was some problem with rim separation and some Shelbys came with the Magnum 500's used on the Boss cars.

Inside, Shelbys followed the pattern of previous years, resembling deluxe Mustangs with Shelby nameplates. This followed for 1969. Shelbys did get a 140-mph speedometer and 8000-rpm tachometer.

Power-wise, it was Mustang city. The 290-hp 351 Windsor was the motivation for the GT-350, and the Cobra Jet 428 for the GT-500. The KR designation never returned after 1968. The 351 did get an aluminum medium-rise intake as did the Cobra Jet.

With Mustang offering its own performance models, including ones more capable of high performance than the Shelby, demand dropped off with production falling to 3,150. The cars were now made at Ford's own Southfield, Michigan, plant.

There were 601 leftover 1969 Shelbys that were updated into 1970 models. Hood stripes and front spoilers were added. Serial numbers were changed and with virtually no promotion, they were sold, marking the end of the Shelby nameplate.

It cost Ford a lot of money to keep the Shelby image cars going, and the return was small. In late 1969, Ford's priorities were rapidly changing and Shelby was an early casualty of a drive to vacate the performance field.

1970

After all it went through in the 1969 model year, Mustangs only had minor changes for 1970, in style, models and engines. Simplification was generally the theme in styling. Grilles were widened with single headlights at the extremities. The outer headlights were eliminated and replaced with a pair of fake vents on each side. A simple rectangular-pattern plastic grille was, with the Mustang symbol, moved from the left side to the center.

Front side-marker lights were moved from the splash pan to front fender be-

Returning for its second and final model year was the 1970 Boss 302. It got the revised Mustang front end for that year and new striping. This preproduction model did not have the correct striping, as it continued behind the rear wheel on production models. It did have the optional Sports Slats and rear spoiler.

hind the vents. Sides were cleaned up with the elimination of the phony vents on the rear quarter panels. Taillights were now indented in twin housings. Inside were minor changes and high-back bucket seats were now standard on all models.

Lineups started the 1970 model year the way they ended the previous season: base hardtop, SportsRoof and convertible models, plus the Mach 1 SportsRoof and Grande hardtop. Back again were the Boss 302 and Boss 429, which continued to be listed as options for the SportsRoof. The GT package was dropped after the 1969 run.

On the surface, engines looked similar to the 1969 offerings, but there were changes. Still standard for all but the Mach 1 and Boss options was the 200-cube six. Optional with the above exceptions was the 250-cid six and 220-hp 302 V-8.

For the second year, 351 V-8's were listed for Mustangs, but for 1970 there were changes. In 1969, all 351's were Windsors. For 1970, the 250-hp units were the only Windsors offered. To complicate it, some 351's were also the new Cleveland block with canted valves, but still carried a two-barrel carb and 250-hp rating.

To confuse the situation still further, there were two variations of two-barrel 351's, with and without Ram-Air, which meant a hood scoop. The non-Ram-Air 351 came standard on the Mach 1.

Four-barrel 351's were now Cleveland units and they too came with and without Ram-Air. Rating was 300 hp at 5400 rpm in either case. Compression was 11:1. The canted valves meant better breathing and were similar in concept to the Boss 302. However, the 351-4V engine used in the 1970 models was fairly calm with two-bolt mains (some got four-bolts) and hydraulic lifters. Hotter versions would be along later.

Back for its final appearance came the 428 Cobra Jet, still rated at 335 hp, be it with or without Ram-Air. Also available was a Drag Pack option with capscrew connecting rods, beefed-up crankshaft, modified flywheel, a vibration damper and oil cooler. It came with a 3.91:1 Traction-Lok or 4.30:1 Detroit Locker differential and could be had with either four-speed or automatic gearboxes. Its rating didn't change.

Boss 302 power again came from the wild 302, but intake-valve size was reduced to 2.19 inches from 2.23 to give a little better bottom-end performance. Aluminum valve covers replaced the chrome ones used the year before.

Actually, Trans-Am cars no longer had to start with five-liter engines for the 1970 model year. A change in the rules permitted destroking larger engines to meet the limit. The move was meant to bring more makes into the Trans-Am, as it was realized not everyone wanted to tool-up for a special racing 302-or-so-sized engine.

Chevrolet responded to the new rule with an LT-1-based 350 for its 1970½ Camaro Z-28. Chrysler went Trans-Am racing for the first time since 1966 and brought out a couple of special models based on the 340, the Plymouth Barracuda AAR 'Cuda and Dodge Challenger T/A. Pontiac destroked its 400-inch V-8. Like Ford, Mercury stayed with the Boss 302 engine, which was available on the Cougar Eliminator.

Returning for a short stay was the meanest Ford engine of them all for 1970, the Boss 429 V-8. The 820T motor was the only one used and for 1970 it only came with mechanical lifters. An Autolite rev limiter, which annoyed the Boss 302 owners in 1969, did the same for Boss 429 owners of 1970 models. Engines came with aluminum valve covers for 1970, while the 1969 models had magnesium and aluminum covers in different parts of the run.

Transmission selection paralleled 1969, but the way to shift the four-speed boxes changed. Hurst shifters and handles were standard. Linkage remained Ford however.

A divided grille with a small one in the middle was the main styling change of the 1970 Cougar, shown here in XR-7 convertible form.

A closer look at the performance models is in order, starting with the Mach 1, which had several detail changes.

Mach 1 grilles picked up rectangular running lights next to the headlights. These were in addition to the directional signal/parking lights below the grille.

Hoods came standard with the fake scoop, with shakers optional. Hoods were now fastened by round latches. The area of flat-black was reduced to a stripe down the middle, slightly wider than the scoop. It ended at the back of the hood instead of spilling over onto the cowl. Next to the big stripe were small ones on each side, which broke next to the scoop for the numbers of the engine size.

Side trim was done with a wide, ribbed-aluminum panel between the wheels which covered the rocker panel and went up onto the bottom part of the door. It was painted flat-black with bright ribs, and Mach 1 lettering was cast into the piece on the front fender. Striping on the spoiler broke for chrome lettering, of similar style to the front fender plate. The rear panel between the taillights got a honeycomb covering in black. The spoiler stripe and striped area on the hood was done in white on some models with contrasting colors.

Instead of the styled steel wheels, Mach 1 got wheel covers that were supposed to look like mag wheels for 1970. They didn't look as good. Standard tires were E70×14 belted.

Interiors were sharpened in detail. Suspensions were toughened-up a bit with a larger front stabilizer bar and a new one in the rear.

Of the Bosses, the 429 was the least promoted and the shortest lived. The fact Ford lost money on them was a factor in production ending at Kar Kraft in December of 1969 after 498 of the 1970 models were built. Early 1970 literature and advertising carried only a passing mention of the 429. "Coax your dealer," was the short message. Mid-year revisions dropped most Boss 429 references.

Unlike the early 1970 Shelbys, the 1970 Boss 429's were not just the earlier models dubbed as 1970's. There were several detail changes. They used 1970 bodies which meant the new grille and taillight treatment and lack of a side scoop. The manual hood scoop returned, but was now painted black instead of the car color as on the 1969 models. Rear stabilizer bars changed from the 1969 type mounted under the rear axle to a design which went over the rear axle. Colors and interior availability also changed for 1970.

Compared to the 429, Ford was generous in its promotion of the 1970 Boss 302. It was featured in literature and advertising. Production increased 331 percent from the partial model year run of 1969, with 7,013 1970 models being built.

Changes included new hood and side stripes. The hood now had a pattern similar to, but not exactly like, the Mach 1. The outer stripes cut to the sides of the hood at the back at a right angle, and overlapped onto the fender top. Below that the Boss 302 lettering was mounted on the fender with the Boss letters in tape and the 302 reversed out of the stripe, which angled down to about a fourth of the height of the side, then shot back to the rear of the car.

Hubcaps and trim rings were now standard with Magnum 500 wheels optional. A rear stabilizer bar was added and was similar to that on the 1970 Boss 429.

A mid-year addition was special trim for the standard SportsRoof, the Mustang Grabber. Teamed with the compact Maverick Grabber, it featured side stripes similar to the 1969 Boss 302 and racing mirrors. No performance car, it came with the 220-hp 302 standard. A later version revised the stripes closer to 1970 configuration, but starting on the door. Engine sizes were worked into the stripes.

Trim changes highlighted the 1970 Mustang Mach 1. New ribbed rocker panel covering carried a cast nameplate. In back, striping changed with letters spelling out the name. Wheel covers replaced the styled steel wheels used in the past.

Mustangs had their toughest competition ever in the sales arena of 1970. There were several new entries: Camaro and Firebird arrived as 1970½ models and only in coupe form, with convertibles being axed. Plymouth Barracuda was all-new in notchback hardtop and convertible models and Dodge got the honor of getting the last pony car, the Challenger, which mirrored Barracuda. The new Chrysler entries could be had with everything from a slant six to the big-horse Hemi and 440. While they were no threat to Mustang in the sales race, they quickly proved to be a major force in drag racing.

A combination of the competition and the shrinking market for performance cars hurt Mustang considerably. It suffered its biggest drop percentage-wise in its history, production was down more than thirty-six percent to 190,727. Unfortunately it was not a reflection of the industry, as domestic production climbed ten percent. Camaro, in a partial model-year-run built 124,889.

Ford's racing budget was cut for 1970 and that meant only one Trans-Am team. Bud Moore, with the potent pair of Parnelli Jones and George Follmer behind the wheel, got the nod and the Shelby Trans-Am cars did not return. Shelby's relationship with Ford was fading at that point anyway.

Factory-backed competition would be at an all-time high for the 1970 season with new teams from Plymouth (Dan Gurney), Dodge (Sam Posey), American Motors (owner Roger Penske and driver Mark Donohue) and Chevrolet (Jim Hall).

However that wasn't all bad, as other teams had a lot of sorting out to do, especially Penske with the Javelin. The Moore crew was coming back with a year's experience and a familiar car.

As happened in 1969, Jones won the season opener, this time at Laguna Seca near Monterey, California, on April 18. Following form of the year before, the second race also went to Mustang; in this case, the May 9 Lime Rock contest, which also was annexed by Jones. However, the Jones car dropped out in the next event May 31 at Bryar Motorsports Park in Loudon, New Hampshire. Follmer was there to take the win and Mustang had three straight. Jones stretched that to four on June 7 at Mid-Ohio, and Mustang had a commanding point lead, thirty-six to sixteen for Javelin and thirteen for Camaro.

Mustang also started well in 1969, but then things fell apart and Donohue got the Camaro working, so Ford followers were apprehensive at Bridgehampton, New York, on June 21 when Donohue drove his Javelin to its first win in three seasons on the Trans-Am circuit. Follmer and Jones took second and third to keep the points tight.

A surprise was in store for the factory teams at Donnybrooke in Brainerd, Minnesota, on July 5 when Milt Minter won in a privately entered 1969 ex-Penske Camaro. Follmer finished second after an exchange of paint. Camaro took over second in points with twenty-six, to twenty-five for Javelin. Mustang was safely ahead with forty-eight.

As the Mustang team feared, Penske and Donohue really got the Javelin sorted out and started winning. Donohue scored back-to-back wins July 19 at Road America in Elkhart Lake, Wisconsin, and at Mt. Tremblant Circuit at St. Jovite, Canada. Mustang's lead was cut to thirteen points, 56-43.

Bringing Ford its first manufacturers championship in SCCA Trans-Am racing since 1967 was the team of (from left) driver Parnelli Jones, car owner Bud Moore and driver George Follmer. Jones won five races during the 1970 season and Follmer took another.

As part of its cutback of the racing budget for 1970, Ford only went with one of its previous two Trans-Am teams, Bud Moore Engineering.

Newcomer to the Jim Hall semi-factory-backed Camaro team, Vic Elford, won the next outing on August 16 at Watkins Glen, New York. Donohue was second and Follmer third, bringing Javelin two points closer with two events remaining on the schedule. It was beginning to look shaky for a Mustang championship.

This time defeat wouldn't be snatched from the jaws of victory. Jones got back on track and took the September 20 contest at Kent, Washington, and the finale, the Mission Bell 250 at Riverside, California, October 4. Donohue was second at Kent and third at Riverside, behind Follmer, who gave Mustang a one-two finish.

Points were counted for the best nine finishes out of the eleven events, so Mustang ended up with seventy-two, Javelin fifty-nine, Camaro with forty and the MoPars were a distant fourth and fifth with Challenger getting eighteen markers and Barracuda fifteen.

Parnelli's series-winning car was tested by *Road & Track* after the season and some of the reasons for success were found in the handling—a Boss 302 strong point.

"It is our nature, maybe even our profession, to second guess, and point out ways various people and companies could improve their products. For once, nothing. We don't know how one would build a better Trans-Am car—and Plymouth, Dodge, Chevrolet, Pontiac and AMC are in the same boat," said an unusually humble *Road & Track*.

At the end of the season, Chrysler, Chevrolet and Ford all withdrew their support for Trans-Am competition, leaving only American Motors with a full factory effort. It was determined to win the manufacturer's title, and did; however, for a while Mustang gave 'em a run for it.

With factory backing gone, Moore decided to try the 1971 series as an independent, using newly built 1970 Boss 302's. Independents Jerry Thompson and Tony DeLorenzo bought a couple of 1970 Mustangs from Moore and gave up on Camaro, so Ford still had a chance against the Javelins.

As expected, Donohue won the opener at Lime Rock on May 8, but DeLorenzo garnered second-place points for Mustang. Jones dropped out in a Moore Mustang.

Follmer ran for Moore at Bryar May 31 and won. Moore also was able to field a second car, driven by Peter Gregg, who nailed down second. Javelin's best was Peter Revson in third, so Mustang took over the point lead, 15-13.

At Mid-Ohio, Follmer scored a second victory with Donohue second and Gregg, third. The lead was 24-19 over Javelin and hopes of a battle between Moore's cars and the Javelins grew.

Unfortunately for Mustang and the race fans, it did not turn out that way. Donohue won the next six events in a row; Moore ran out of parts and money and quit; and AMC had its championship with seventy-two points to fifty-four for Mustang.

Ironically, the final event at Riverside on October 3 was taken by Follmer, but not in a Ford. He lined up a ride in American Racing Associates independent Javelin. Donohue had other commitments and Jackie Oliver drove the Penske car to third, behind Vic Elford in another ARA Javelin.

After the 1971 season, AMC also pulled out, to concentrate on NASCAR, but the Bud Moore-built Mustangs were raced for several more seasons in Trans-Am and other sports car events by their new owners. The large 1971 Mustangs never did catch on in sports car racing.

Another area of popularity for 1970 and older Mustangs is in short-track racing. Although the chassis is usually not stock, Mustangs were popular for most of the 1970's in short-track racing with few being the 1971 and newer variety.

White-glove treatment was supposedly reserved for the early 351 Clevelands. Note the large exhaust manifold passages and front timing-chain cover that was cast into the block.

Dick Trickle of Wisconsin Rapids, Wisconsin, set a national record in 1972, winning sixty-seven short-track features in a single season. He drove a 1969 Mustang, which was updated to a 1970 mid-season.

The 1970 drag racing season marked the beginning of pro-stock competition. The idea was to have a professional racing class with stock-appearing bodies and engines from that make of car, though not necessarily that model. While some Mustangs did compete in pro-stock, the 1970 year coincided with introduction of the 103-inch wheelbase Maverick, which got the factory pro-stock backing. Mavericks carried Boss 429's and sohc 427's. Later, when rules permitted, Pintos were used as the basis for pro-stocks, leaving Mustangs in the too-large category, at least through the 1973 models.

Since the drag racing development of the Maverick and Pinto did not translate into performance models for the street, as was the case with Fords in earlier years, those nameplates will not be covered in this volume. However, the success of Fords in pro-stock in the following years would give the cars from Dearborn some of their finest hours in drag racing history.

The timing of pro-stock turned out to be wise, as the new production cars would begin to go downhill in performance, starting with the 1971 models. Since pro-stock engines were far more wide-open, the output of the newest production cars made little difference.

Pro-stock diminished the importance of super stock racing, but the downturn in new car performance was already changing the face of SS racing. In the early and mid-1960's, the hottest new cars squared-off in SS. As the decade progressed and the number of SS classes increased, older models gained in popularity. The fast cars of earlier years still were good performers and SS became a class for the hot older cars, just like the stock classes.

Mustangs, usually 1970 and older were raced in SS competition for years to come. Some examples of the 1971-73 Mustang were raced at the drags, but they were rare compared to the older models. Chevrolets, Chrysler products and Pontiacs were usually more common, mainly because of the limited availability of Ford parts.

Ford was trying to get its Muscle Parts program in high gear when orders came through to cut back performance activities. The phases of the cutbacks were in late 1969, late 1970 and in 1973.

While Chevy and Chrysler never really stopped their performance parts programs, Ford did and as a result, as the 1970's progressed, Ford's parts situation for the average buyer got worse. It wasn't that a Ford couldn't be made to go fast, the pro-stocks and NASCAR stockers proved that when properly set up, Fords could win and win big; it was just that parts were hard to get and were usually expensive.

What this all meant for Mustang was that after the 1970 models, the racing career of the Mustang pretty well ended for the rest of the decade. It wasn't until ten years later when Ford got back into racing and performance that Mustangs, new ones that is, would be potential winners in major stock and sports car events.

A new dashboard was used for the all-new 1971 Mustangs. It had two large dials in front of the driver, plus a small one for fuel. This Mach 1 panel has the optional instrumentation group with a tachometer in the left dial and three smaller gauges in the center.

Several items up front distinguished the 1971 Mach 1 from other Mustangs. The twin-scooped hood was optional at no extra charge on 302 V-8 models and standard on others. The leading edge of the hood and fenders were painted instead of chrome, as was the front bumper which was coated with urethane plastic. The grille was unique with a honeycomb pattern and small emblem, plus driving lights. The Mach 1 decal is on the front fender, behind the wheel opening.

1971

As was the case every two years, Mustang was the recipient of a new styling exercise for 1971. Unfortunately not all the features of the all-new body were improvements. While the 1971 models were introduced at a time when the majority of Ford's remaining racing programs were about to be curtailed, they were conceived during the performance days and intended to be big enough between the front wheels to house any engine Ford could come up with at the time and in future years.

Ford seemed to always get itself into situations where every time it restyled a car, it made it bigger—and the 1971 Mustang turned out to be yet another example.

Wheelbase on the 1971 models grew for the first time, to 109 inches from its original 108. Length popped up 2.1 inches and width 2.4. Front tread grew three inches and rear 2.5, coming to 61.5 inches in the front and a half-inch less in back. There was now room for practically anything under the hood and no need for special assembly lines, as was the case with the Boss 429's.

It didn't take long for the press to turn back a couple of pages and compare the 1971 Mustang with the original 1965 model. Wheelbase gained an inch, length 7.9 inches and width 5.9 inches. Weight also increased about 500 pounds. Ford wasn't alone in its bulk campaign as Mercury Cougar went through the same process and now had a 112-inch wheelbase which was the same as some intermediates.

Despite the changes, the 1971 Mustang wasn't a terrible car. Far from it. Styling was interesting and the selection of performance engines was excellent. There even was a genuine performance model, the Boss 351, plus our old friend, the Mach 1.

Front end styling was reminiscent of the 1969 Shelby with a full-width opening, single headlights at the ends and a grille in the center. On models with chrome bumpers, the bumper was thin and surrounded the lower part of the grille, while chrome trim took care of the rest.

Returning for the first time since the 1967 models were the grille bars with the horse symbol in a corral in the middle. The grille opening itself was surrounded by bright trim and had a honeycomb pattern for a background.

Mach 1, Boss 351 and a mid-year Sports Hardtop came up with their own grille and bumper combination. A plastic-coated front bumper extended out from the grille and was in body color (although some Boss 351's got chrome bumpers). Surround trim for the grille also was done in body color. Grille bars were gone and running lights were mounted in the honeycomb grille. These were in addition to the directional signals in the scooped valance panel below the front bumper. The grille itself was surrounded with bright trim with a small Mustang symbol in the center.

Hood and cowl design changed on the 1971 models. The standard hood was fairly flat and extended from the grille to the windshield, covering the cowl and windshield wipers.

Standard on Mach 1 models with 351 and larger V-8's and optional at no charge on 302-powered Mach 1's was a hood with a raised center section and two NASA-type scoops on either side. These were not functional, but an optional Ram Induction setup for them was, which had plastic ducting to get the fresh air to the carburetor. It was available for cars with 351 or 429 V-8's and standard on the Boss 351. Decals on the side said Ram-Air and the engine size. Round hood pins held it down.

Windshields sloped 5.5 degrees more and now were slanted at a sixty-degree angle. Roofs were lower with SportsRoof models dropping to 50.1 inches from 50.6. Hardtops went from 51.5 to 50.8 and convertibles 51.5 to 50.5.

Most radical of the designs was the SportsRoof with nearly a straight roof line from the passenger compartment to the end of the trunk lid. A large flat rear window provided limited rear vision. Rear quarter window design changed and now was a roll-down unit like on the conventional hardtop, ending the fastback tradition of different rear quarter opening styling.

Two-door hardtops still had formal lines, but the trailing edge sail panels swept back a bit more to the trunk area.

Side sculpturing was simple with a line streaming straight back from close to the grille to the rear quarter. There also was another lower body line, similar to the earlier models. Rear quarter kick-up was more pronounced.

Rear styling carried the familiar Mustang theme with triple-lens taillights on each side and gas filler in the middle. Taillights were rounded on the sides and now contained the backup light in the center lens, eliminating the separate backup lenses in the valance panel.

Interiors were redone with a new plastic dashboard. Two large round dials and a small one between them faced the driver. Standard was a bank of warning lights in the left, the fuel gauge in the center and speedometer in the right. The optional instrumentation group included an 8000-rpm tachometer in the left dial and the addition of a panel of three gauges in the center section of the instrument panel, and tilted toward the driver. In the panel were oil pressure, ammeter and temperature gauges. The group was standard on the Boss 351 but not available with the six-cylinder engine.

Jutting outward a couple of inches was the raised center section of the dash. It contained air-conditioning outlets, a place for the optional gauges (there was a plaque in its place with the standard instruments), radio and heater controls. The right side was a large, flat area with the glove compartment on the lower half. Manufacturers didn't know if they would have to have a place for air bags, so several dashboards of the era contained large expanses of unused space.

A mini-console was standard, surrounding the shift lever on the floor. The optional console blended in below the dash and contained an electric clock, armrest and storage area. High-back bucket seats were standard.

Engine availability included several changes. Standard, except for the Mach 1 and Boss 351, was a larger 250-cube, 145-hp six in place of the old 200-cid unit. Standard on the Mach 1 and optional for all models except the Boss was the 302 V-8, rated at 210 hp.

Back again came the 351 Windsor, in two-barrel, 9:1, 240-hp form. Ram-Air was optional, but did not change the ratings.

Startling to live up to its performance potential was the351 Cleveland. In four-barrel configuration, it had a 10.7:1 compression ratio and advertised output of 285 hp at 5400 rpm. It also came with optional Ram-Air and unchanged ratings.

Generally considered the best of the production 351's was the Boss HO engine. It started with a four-bolt underside and high nodular crankshaft. Rods and pistons were beefed up, there were solid lifters working on a high-lift cam, an aluminum intake and Autolite 750-cfm four-barrel. Heads were very similar to the Boss 302 with 2.195 intake and 1.714 exhaust valves, but had detail features for durability, and were highly prized by 351 engine builders. Compression ratio averaged 11:1. Advertised horsepower came to 330 at 5400 rpm and torque was 370 pounds-feet at 4000.

Putting the 351 into perspective, 1971 was the model year that General Motors announced that all of its passenger cars would run on low-lead or no-lead fuel, meaning compression ratio cuts for all of its high-performance engines, including those on the Corvette. Chrysler, American Motors and Ford all said they too would be cutting their compression ratios in the future, but not at the rate at which GM tackled the project.

The cutbacks weren't just to hamper the performance buyers but rather to help meet the rapidly stiffening emission regulations. Ford introduced its low-compression performance engine as an option for the Mustang in May of 1971, the 351-4V CJ. It was available for all models except the Boss 351. "The introduction of the 351 CJ recognizes the trend of performance buyers toward lighter weight,

There was only one Boss Mustang for 1971, the Boss 351. Using a SportsRoof body, it came with a 330-hp 351 HO standard. Exterior goodies included a Mach 1 grille, Ram-Air hood with blacked-out surface and round hood fasteners. Side trim included stripes, decal on the front fender behind the wheel and lower blacked-out area. This model had a chrome front bumper, others had plastic coated front bumpers. The front spoiler was standard, as were the flat hubcaps, wheel trim rings and raised-letter F60x15 tires.

for better handling, and lower displacement engines," said John B. Naughton, Ford Division general manager.

It retained the four-bolt mains, but used hydraulic lifters. An aluminum intake manifold featured a new Autolite 750-cfm four-barrel with new spread-bore design, with larger secondary areas. Compression dropped to 8.6:1 thanks to different pistons and open chamber head design. Advertised output was 280 hp at 5800, while torque came out to 345 pounds-feet at 3800, with or without Ram-Air.

One of the reasons for the wider front stance was to provide room for the 429-cube engines, Boss or otherwise. Earlier designs always had problems accommodating larger engines.

A couple of Boss 429's were built using the 1971 body, but the production setup was the conventional 429 performance engines, like those used in Torinos. You could get the 429 CJ with or without Dual Ram Induction. In the former condition it got the name 429 CJ-R. Either way the horsepower rating was advertised as 370 at 5400 rpm.

The 429-equipped Mustangs did little to solve the handling problems of earlier models with big blocks. "With all that beef on the front, the fact of the matter is great smoking understeer," said *Sports Car Graphic*. It got a 6.3-second 0-60 and quarter in the four-speed, 3.5-geared car of 14.6 seconds and 99.4 mph.

A third version was the Drag Pack, which was similar to the one on the Torino. It came with a choice of locking differentials and the Super Cobra Jet 429. Buyers had a choice of 3.91 Traction-Lok or 4.11 Detroit Locker gearing.

The Super Cobra Jet was considerably different from the 370-hp CJ. It had four-bolt mains, not two, mechanical lifters instead of hydraulic and other features like capscrew rods, high-lift cam and modified crankshaft. It came with the fresh-air hood and carried an advertised rating of 375 hp at 5600 rpm. Torque was shared with the CJ at 450 at 3400. Compression also was the same at 11.3:1.

Model lineup for 1971 included base hardtop, SportsRoof and convertible, Grande hardtop and Mach 1. The Boss 351 technically was an option. Mid-year a Spring Special Value option brought a Mach 1 grille, hood and Boss 351 striping and paint to the regular hardtop.

In addition to the features already described, the Mach 1 came with flat-black or argent lower fender and rocker panel trim, topped by a narrow chrome strip, a large Mach 1 decal on the front fender, a Mach 1 decal and striping on the rear of the deck lid and honeycomb pattern applied to the area between the taillights, similar to that of the grille.

A standard SportsRoof interior was used, but a special Sports Interior was optional for the Mach 1. Cars with dual exhausts got chrome extensions, which exited below a modified rear valance panel. Flat hubcaps and trim rings were again standard, with F70x14 white-stripe tires. Under it all again came the Competition Suspension.

Boss 351's came with four-speed manual boxes only, which again had as standard equipment the Hurst shifter. Also in the package came the Competition Suspension, 3.91 Traction-Lok, F60x15 belted white-letter tires, the rpm limiter and dual exhausts which ended before the valance panel.

A Mach 1 grille, front spoiler below the front valance panel, Ram-Air hood, lower body black-out, special side stripes which started at the front side-marker light then went up and turned at a right angle for the back, tapering along the way (the stripes were optional for the Mach 1), Boss 351 decals on the front fenders and rear deck, and blacked-out rear panel all distinguished the Boss 351. Interior was like the SportsRoof, with all options, but did include the Instrumentation Group.

Also going through a balloon stage in its career was the Mercury Cougar, which for 1971 sat on a 112-inch wheelbase. This is the XR-7 two-door hardtop.

Considering all the parts, the Boss 351 was a well-equipped, well-performing package. However, it lacked the magic and racing heritage of the Boss 429 and 302, and didn't do especially well in the showroom or in print. It was considered an option and not listed in year-end production figures. Estimates put the total just over the 1,800 mark.

For some reason, Ford modified the cars that went to the car-buff books for testing. Certainly the most favorable acceleration figures had to be a goal. *Hot Rod* did the quarter in 14.09 seconds with a trap speed of 102.73 mph, which it noted was faster than the 1970 Boss 302. The testers also noted it weighed 150 pounds more and didn't handle as well.

Also running a modified Boss 351 was *Sports Car Graphic*, which reported a 14.7/96.2 quarter and 0-60 sprint of 6.6 seconds. "For being bigger, the Mustang is still a very nimble-feeling machine."

While the car was bigger, sales were not and Mustang's slide continued. Only 149,678 were called for, making a 21.5-percent drop in demand. Things weren't much better in the other pony car camps. Camaro was down to 114,643, to hold down second in the class.

1972

Rumors of impending demise for the pony car category circulated, but Mustang was back for more, with less, for 1972. For the new model year the list of what didn't make it was more interesting than what did. Among the departed were the Boss 351 and its engine, all of the 429's and any remaining thirst for premium fuel.

Starting with the 1972 models, net horsepower and torque figures were used instead of the old gross system. It was just as well, for comparisons with the previous figures with high-compression engines would have proved embarrassing.

Standard, except for the Mach 1 came the 250-cube six, now at 99 hp. Standard on Mach 1 and optional on other models was the 302, billed at 141 hp. The 351 Windsor carried a 177-hp reading and the four-barrel 351 Cleveland was demoted to 266 hp and had the same 8.6:1 compression as the Windsor. Some sources call the latter engine the CJ. It was very close to the 1971 351 CJ.

The hottest engine for 1972 was available only in limited numbers, the 351 HO (High Output). For all practical purposes, it was a reincarnation of the high-compression 351 HO in the 1971 Boss 351 but with lower compression thanks to open chamber heads and flat-top pistons, resulting in a 9.2:1 squeeze.

Advertised HO output was 275 hp at a healthy 6000 rpm. It would be Ford's last domestic performance engine for the decade.

Even though there was no special car for the 351 HO this time around, if you were lucky enough to get one installed, you also had to have the four-speed wide-ratio transmission (the close ratio was history), 3.91 Traction-Lok differential, Competition Suspension, rpm limiter, power front discs and F60x15 belted tires.

Sound familiar? It should, it was the list of standard equipment for the 1971 Boss 351. Checking the HO box on the order form wasn't especially economical, as the package with the above and a few other goodies cost $894 over the 302 V-8; unless you were ordering a Mach 1 which had some of the stuff standard and reduced the bill to $865.

The cars themselves had only minor detail changes for 1972 with slightly changed trim. Mid-year the Ram-Air hood was only available for the 351-2V Windsor engine, which didn't make much sense.

Ford was more interested mid-year in promoting its Sprint Option packages (there were two kinds) for the hardtop and SportsRoof. Special red, white and blue interiors and exteriors were featured; but as with previous Sprints, performance was not a consideration.

One remaining vestige of performance was the Hurst shifter, which came with four-speeds, which now could be bolted to only the 351-4V or 351 HO. A mere

One of the few alterations for 1972 Mustangs was the change in rear deck ornamentation. Instead of individual letters across the back, a Mustang script nameplate was on the right-hand side. The feature continued for 1973.

2.7 percent of the 1972 Mustangs were so equipped, the lowest in the history of the nameplate. The total had been as high as 14.5 percent in the 1965 models.

The final performance Mustang of the first generation was still fairly agile. *Car and Driver* hopped in a four-speed HO-equipped SportsRoof and found sixty in 6.6 seconds and the end of the quarter mile in 15.1 at 94 mph.

New car production for the 1972 model year climbed nearly nineteen percent, but guess who didn't get in on the boom? Mustang hit bottom with 125,093 being built for a 16.4 percent drop.

To show how far things had declined, a look back at the record model year of 1966 reveals 607,568 Mustangs rolling down the assembly lines. By the 1972 run, 79.4 percent of those orders were gone. However, Mustang's bad times were nothing compared to Camaro. A 174-day strike at the factory which built them cut production to 68,656.

1973

Surprisingly the first Mustang to reverse the sales slide was the first to break the two-year styling cycle, the 1973 Mustang. Though there were only minor changes, its production increased to 134,867 for a 7.8 percent gain.

A factor in the small increase was word that this would be the last of the big Mustangs. The 1974 models were to be smaller and based on Pinto underpinnings. Also the 1973 would be the last Ford and Mustang convertible (at least that's what was thought then) and increased popularity of the convertible accounted for 55.8 percent of the production increase. Convertible production at 11,853, was up considerably over the 6,401 run for the 1972 models, had the best numbers since 1969.

Camaro production also rebounded, but was still well below Mustang at 96,756, making the first-generation Mustangs undefeated in the pony car sales sweepstakes.

The last of the first generation Mustangs suffered another drop in performance with the 351 HO being discontinued and the horsepower ratings of most engines lowered as stricter emission standards took their toll. The six now was a 95-hp weakling, the 302 V-8 was still at 141 hp and the 351W at.173 hp. The Windsor engine was the only one for which the Ram-Air option was available, as was the case since mid-1972.

There is a great amount of discrepancy in the 1973 horsepower figures. Ford no longer advertised them in its literature, and to make matters worse, said that the output varied per car, depending on factors such as model, equipment, etc. The system was probably correct, but for the folks who kept track of such things, it drove them crazy. Figures used here, except as noted, were taken from Ford's report to the Automobile Manufacturers Association, as revised in December of 1972.

Ford's most powerful engine for the Mustang, or for any car in 1973 for that matter, was the 351 CJ. Its AMA rating was 259 hp at 5600 rpm, but other ratings used indicated 246 and 248 hp at 5400. Compression was 7.9:1 compared to 8:1 for the other engines and marked yet another drop. Dish-topped pistons were now needed. Inside, the CJ still had the good stuff. It was the only engine which could be had with the four-speed. Gearing was a choice of 3.5 or 3.25, whether the four-speed or automatic was used.

Mach 1 for 1973 featured changed side striping, a new grille design with the directional signals in it instead of in the valance panel and scooped hood for most models. This copy has the optional forged aluminum wheels. Rear quarter windows were fixed, unless the optional power-window controls were ordered.

New federal bumper standards took effect for 1973 models and to meet them, a larger, urethane-coated front bumper and extended and remounted rear bumper brought the overall length to 193.8 inches, an increase of 4.3 inches.

The larger front bumper all but hid the front valance panel, so the directional signals were moved up to the restyled grille. Headlights were in square bezels at the ends. A rectangular center section had the vertical directional signals at the ends and different patterns in the center. A large egg-crate plastic center portion with the Mustang symbol in its corral, mounted by short vertical bars, was on base models.

Mach 1 and cars with the optional Exterior Decor Group had a pattern that looked like expanded metal and the smaller Mustang emblem. Insides of the headlight bezels were blacked-out instead of having brightwork. All cars had painted trim surrounding the grilles. The Decor Group again also contained black or argent lower body paint and chrome strip border.

New tape stripes made up the side trim for the Mach 1. The front portion started behind the bumper, broke for the wheel opening and continued on the other side, ending before the rear wheel opening. The Mustang script chrome plate was in the middle of the strip on the front fender and Mach 1 lettering was reversed out on the rear quarter. It was located about a third of the way up the side.

New rear striping over and around the taillights had Mach 1 lettering on the right-hand side. The rear panel between the taillights again had a honeycomb pattern.

Other Mach 1 features continued to be the Competition Suspension, the scooped hood (you could have a Mach 1 with or without at no charge), a standard 302 V-8 and E70x14 belted white-stripe tires.

Technically, the Mach 1 and regular SportsRoof were no longer two-door hardtops, at least in standard form. Starting with the 1972 models, the rear quarter windows were fixed, except when power windows were ordered, with which they retracted. An interesting option for 1973 was forged aluminum wheels, which replaced the Magnum 500 units on the option list.

1974-1978

The 1973 models were the end of the line for the first-generation Mustang. Starting with the 1974 models, Mustang would be a smaller car with its heritage in the subcompact Pinto line rather than in the Falcon line as was the case with the first Mustangs. Wheelbase was 96.2 inches and at first you couldn't even get a V-8.

The timing couldn't have been better as far as Ford was concerned, for the introduction coincided with the start of the so-called energy crisis, when oil companies got to boost their prices and cut their output with the government cheering them on.

Though there were attempts to create some sort of a mini performance image for the new Mustang II, including revival of the Cobra name, the 1974-78 Mustang II's were not performance cars and most Mustang and performance car fans today recognize that.

1979-1987

The third-generation Mustang (it wasn't called the Mustang III) arrived for the 1979 model year and was based on a shortened Fairmont chassis. Wheelbase grew a bit to 100.4 inches. At the same time, Mercury transferred its Capri name from a series of German imports to a cousin of the new Mustang. It marked the first time since 1973 (with the Cougar) that Mercury had an entry in the sporty car field.

The performance version of the 1979 Ford Mustang was the Cobra option, which included a turbo four, heavy-duty suspension and, of course, a hood decal.

A mid-year addition to the Mustang line in 1979 was the Limited Edition replica of the Indianapolis 500 pace car. It came with turbo four or mild 302 V-8 power.

The Mustang came in notchback and hatchback models, and the Capri only in hatchback form. Of course, all were two-doors.

The closest thing to a performance version of the early third generation Mustangs and Capris was the option of a turbocharged four. It was part of the Cobra package on the Mustang and Turbo RS option for the Capri. This setup continued through the 1981 model year, during which the turbo was dropped, because of various durability and mechanical problems.

A significant step in the return of the Mustang to serious performance came in the mid-1979 Limited Edition, which was a pace car replica for the Indianapolis 500. While the cars on the street had the meek 140- horse 302 V-8 that brought complaints in the Mustang II, development by Ford engineers for the cars used at the track seemed excessive—and was excessively promoted. Old parts bins were raided and much was made of Ford's performance history.

A few months later, Ford gradually let the string unwind and announced racing contingency programs, followed by the setting up of its Special Vehicle Operations (SVO) department for racing and hot production cars.

It got worse before it got better: The top V-8 for the Mustang and Capri dropped to the lethargic 255 for 1980 and 1981, before the introduction of the Mustang GT for 1982. That came with a 157-horse 302 V-8, which would only get healthier.

For 1983, the GT went up to 175 net horsepower. Also, a Turbo GT with four-cylinder power came on-stream for 1983, as Ford powers were torn between V-8 and four configurations as the performance powerplant of the future.

To enhance Mustang, a convertible was added to the lineup mid-year in 1983. From the start, it was available as a GT. No Capri version was forthcoming, although specialty-car-converter American Sunroof Corporation offered its Capri ASC McLaren soft top and hatchback in 1985 and 1986.

For 1984, the GT V-8 started at 175 horses, but the addition of roller rockers and better exhaust headers upped it to 205 a couple of months into the model run. You could also get GT and Turbo GT convertibles.

A special version of the hatchback, the SVO, bowed for 1984 and featured a turbocharged, fuel-injected version of the 140-cube overhead cam four, rated at 175 horses, compared to the 145 in the Turbo GT. Special aerodynamic bodywork was part of the SVO package.

Buyers voted with their dollars and it was hands down, the Mustang V-8's won. The Turbo GT left the lineup for 1985, and the towel was thrown in on the SVO before the 1986 model chase ended, with production well below original estimates.

Meanwhile, the GT gained five more horses for 1985.

An attempt at sophisticated turbo four power and sports mechanicals was the 1984 Mustang SVO. It featured a different nose than the rest of the line. It was not a success, despite a facelift in 1985.

The first serious attempt at making Mustang a performance car again was the 1982 GT, which came with a high-output 302 V-8 engine. Adding to the rejuvenation was a convertible, which joined the line-up in mid-1983. It came as calm V-6 GLX or muscular GT (shown).

Mercury Capri shadowed the Ford Mustang from 1979 through 1986, and got most of the mechanical and styling goodies its cousin did. This is the 1982 Capri RS with the 302-cube, 157-horse V-8 and appropriate external gingerbread.

For 1986, electronic fuel injection replaced the four-barrel carburetor on the Mustang GT, but advertised output dropped a score, to 200. However, in nearly every comparison with rival Chevrolet Camaro IROC-Z, the GT won the performance category rather easily, prompting Chevrolet and Pontiac to go to a 350-cube V-8 for 1987, in addition to the old 305.

Capri engines paralleled Mustang for most years, except there was no SVO counterpart. The RS option was more or less the equivalent of the Mustang GT. For 1986, top performance Capris were designated 5.OL.

The Capri was dropped from the Mercury lineup for 1987, as the division preferred to concentrate on its Merkur XR4Ti, which was imported from Germany and carried turbo four power.

For 1987 the Mustang got its most extensive facelift since it was all new for 1979. Aerodynamic noses with flush headlights, flush side glass and restyled rears with trick taillights were among the features. Also the dashboard got a major update.

A 225-horsepower version of the injected 302 V-8 was ready for the Camaro and Firebird 350's. The V-6 engine was dropped, leaving only the base four and muscular 302. If necessary, an HO version of Ford's 351 V-8 was waiting in the wings.

The extensive changes mean that the Mustang, in its current rear-drive form, has a couple of years to go in the increasingly front-wheel-drive performance-car market, but rumors of replacements continue, most around a car developed with Mazda of Japan.

Mustang and Capri racing has been concentrated on sedan sports car racing events, like the SCCA Trans-Am and IMSA Camel GT GTO series. Factory-backed teams of Mustangs and Capris dominated their classes in recent seasons, although Merkurs started to replace the Capris in Trans-Am racing during 1986.

It looked hopeless for a while, but Mustang has returned to its prominent status in the world of performance cars. With the current kick by Ford, it looks as if it will stay that way in the foreseeable future.

For collector value, the 1965-70 Mustangs reign supreme, the 1971-73 models are making some inroads, the 1974-78's will probably never rival the older cars, and the 1979-87's have a long way to go before a direction can be seen—but certainly in some forms, they look good.

Look closely and you will see this is a Mercury Capri, used in SCCA Trans-Am racing. Both Mustang and Capri body parts were shifted when necessary. The chassis of the cars were far from stock. This is Chris Kneifel driving a Trans-Am Capri in 1985 at Road America.

New front and rear styling, plus the full aero treatment mark the 1987 Mustang GT convertible. Horsepower has been boosted to 225, thanks to new cylinder heads.

A new nose, flush side windows, full aero skirting and a stronger 225-hp 302 V-8 are among the features of the 1987 Mustang GT, which is again a front-running performance car.

SPECIFICATIONS

Following is an abbreviated guide to the 1959-1973 Fords for which there was some performance potential, performance models or continuation of a performance name; models such as wagons and 4-door sedans are not included.

While all the specifications cannot be listed, ones with some bearing on performance, sporty looks or equipment are included. Information was taken from several sources and in some cases, judgements had to be made when there was conflicting data.

Also, when dealing with hundreds of thousands of cars, it is impossible to make them all fit in simple categories. While mid-year changes have been noted wherever possible, there are certain to be exceptions to the general outlines here. It should be remembered that these specifications are intended to apply to the majority of cars Ford built.

One possible area of confusion is in the model numbers listed. Actually these are combinations of two-digit numbers and letters. They were called several things including body code, body style number and model number. They are found listed on the warranty plate and/or certification label under the space listed for the body.

Ford added further to the situation by sometimes assigning special trim options their own body code, especially on the 1965-70 Mustangs, and then not assigning separate body codes to performance models like the Torino Talladega and Cobra and Boss Mustangs.

Model numbers listed are for the base models, without options.

Engine codes listed appear in the vehicle identification number. The first number of the VIN is the last number of the model year; next comes a letter telling what plant the car was built in; then a two-digit number for body serial (or sales code) that does not break down to include series; following is a single letter for the engine code; finally is the consecutive unit number.

Engine specs and ratings are as advertised. Sometimes running changes were made in output and compression. They are also listed in many cases.

Equipment on performance models or in performance options is listed for most items, but some minor items not directly related to performance or its image are omitted.

Emblems for various engines are listed for the majority of cars but, especially in this area, there were exceptions in many cases, sometimes varying from factory to factory.

Option lists are limited to items that pertain to performance and/or sporty trim.

Both new car and option prices varied in some years, being changed several times. In years of frequent changing, the time at which the prices were effective is noted.

ABBREVIATIONS

WB	Wheelbase
OL	Overall length
OW	Overall width
FT	Front tread width
RT	Rear tread width
HT	Overall height with design load
CID	Cubic-inch displacement
BxS	Bore and stroke of engine
Hp	Advertised horsepower
Tq	Advertised torque in foot-pounds
CR	Compression ratio
Carb	Carburetor
Std.	Standard
Opt.	Optional
NA	Information not available

1959 THUNDERBIRD

DIMENSIONS

WB: 113"
OL: 205.4
OW: 77
FT: 60
RT: 57
HT: 52.5 (hardtop) 53.1 (convertible)

MODELS

Series	Body Style	Model No.	Price
Thunderbird	2-dr hardtop	63A	$3696.
Thunderbird	2-dr convertible	76A	3979.

ENGINES

Code	Type	CID	BxS	Hp@Rpm	Tq@Rpm	CR	Carb
H	V-8	352	4x3.5	300@4600	380@2800	9.6	4-bbl
L	V-8	430	4.3x3.7	350@4400	490@2800	10	4-bbl

TRANSMISSIONS

3-Speed (Std. 352/300)
Overdrive (Opt. 352/300)
Cruise-O-Matic (Std. 530/350, opt. 352/300)

SUSPENSION

Front: Conventional coils and stabilizer bar
Rear: Leafs

OPTIONS

430/350 V-8, $177; Overdrive, $108.40 (352/300 only); Cruise-O-Matic, $230.80.

1960 THUNDERBIRD

DIMENSIONS

WB: 113"
OL: 205.4
OW: 77
FT: 60
RT: 57
HT: 52.5 (hardtop) 53.1 (convertible)

MODELS

Series	Body Style	Model No.	Price
Thunderbird	2-dr hardtop	63A	$3755
Thunderbird	2-dr convertible	76A	4222

ENGINES

Code	Type	CID	BxS	Hp@Rpm	Tq@Rpm	CR	Carb
Y	V-8	352	4x3.5	300@4600	380@2800	9.6	4-bbl
M	V-8	430	4.3x3.7	350@4400	490@2800	10.0	4-bbl

TRANSMISSIONS

3-Speed (Std. 352/300)
Overdrive (Opt. 352/300)
Cruise-O-Matic (Std. 430/350, opt. 352/300)

SUSPENSION

Front: Conventional coils and stabilizer bar
Rear: Leafs

OPTIONS

430/350 V-8, $177; Overdrive, $144.50 (352/300 only); Cruise-O-Matic, $242; Sunroof, $212.40.

1961 THUNDERBIRD

DIMENSIONS

WB: 113"
OL: 205
OW: 75.9
FT: 61
RT: 60
HT: 52.5 (hardtop) 53.3 (convertible)

MODELS

Series	Body Style	Model No.	Price
Thunderbird	2-dr hardtop	63A	$4170
Thunderbird	2-dr convertible	76A	4637

ENGINES

Code	Type	CID	BxS	Hp@Rpm	Tq@Rpm	CR	Carb
Z	V-8	390	4.05x3.78	300@4600	427@2800	9.6	4-bbl

TRANSMISSIONS

Cruise-O-Matic

SUSPENSION

Front: Conventional coils and stabilizer bar
Rear: Leafs

1962 THUNDERBIRD

DIMENSIONS

WB: 113"
OL: 205
OW: 76
FT: 61
RT: 60
HT: 52.5 (hardtop) 53.3 (convertible)

MODELS

Series	Body Style	Model No.	Price
Thunderbird	2-dr hardtop	63A	$4321
Thunderbird	2-dr convertible	76A	4788
Landau	2-dr hardtop	63B	4398
Sports Roadster	2-dr convertible	76B	5439

ENGINES

Code	Type	CID	BxS	Hp@Rpm	Tq@Rpm	CR	Carb
Z	V-8	390	4.05x3.78	300@4600	427@2800	9.6	4-bbl
M	V-8	390*	4.05x3.78	340@5000	427@3600	10.5	3x2-bbl

*Mid-year engine.

TRANSMISSIONS

Cruise-O-Matic

SUSPENSION

Front: Conventional coils and stabilizer bar
Rear: Leafs

PERFORMANCE MODELS

Sports Roadster—Model included fiberglass tonneau cover with built-in headrests; Kelsey-Hayes wire wheels; passenger grab-bar on dashboard; emblem on front fender under Thunderbird name.

OPTIONS

390/340, $242.10; wire wheels $373.30 (Std. Sports Roadster).

1963 THUNDERBIRD

DIMENSIONS

WB: 113"
OL: 205
OW: 76.5
FT: 61
RT: 60
HT: 52.5 (hardtop) 53.3 (convertible)

MODELS

Series	Body Style	Model No.	Price
Thunderbird	2-dr hardtop	63A	$4445
Thunderbird	2-dr convertible	76A	4912
Landau	2-dr hardtop	63B	4548
Sports Roadster	2-dr convertible	76B	5563

ENGINES

Code	Type	CID	BxS	Hp@Rpm	Tq@Rpm	CR	Carb
Z	V-8	390	4.05x3.78	300@4600	427@2800	9.6	4-bbl
M	V-8	390	4.05x3.78	340@5000	427@3200	10.5	3x2-bbl

TRANSMISSIONS

Cruise-O-Matic

SUSPENSION

Front: Conventional coils and stabilizer bar
Rear: Leafs

PERFORMANCE MODELS

Sports Roadster—Model included fiberglass tonneau cover with built-in headrests; Kelsey-Hayes wire wheels; passenger grab-bar on dashboard; emblem on front fender (on most cars).

OPTIONS

390/340, $242.10; wire wheels $373.30 (Std. Sports Roadster).

1963 FALCON

DIMENSIONS

WB: 109.5"
OL: 181.1, 190 (wagon)
OW: 70.6
FT: 55
RT: 54.5
HT: 54.5 (sedan), 53.2 (hardtop), 53.8 (convertible), 54.9 (wagon)

MODELS

Series	Body Style	Model No.	Price
Falcon	2-dr sedan	62A	$1985
Futura	2-dr sedan	62B	2116
Futura	2-dr sedan (Sports)	62C	2236
Futura	2-dr hardtop*	63B	2198
Futura	2-dr hardtop (Sports)*†	63C	2318
Futura	2-dr convertible	76A	2470
Futura	2-dr convertible (Sports)†	76B	2590

*Mid-year model †Sprint Option available (Mid-year)

ENGINES

Code	Type	CID	BxS	Hp@Rpm	Tq@Rpm	CR	Carb
S	I-6	144	3.5x2.5	85@4400	134@2000	8.7	1-bbl
U	I-6	170	3.5x2.94	101@4400	156@2400	8.7	1-bbl
F	V-8	260*	3.8x2.87	164@4400	258@2200	8.7	2-bbl

*Mid-year engine

TRANSMISSIONS

3-Speed (Std. 144/85, 170/101)
3-Speed All-Synchro (Std. 260/164)
4-Speed (Opt. all)

Fordomatic (Opt. all)

SUSPENSION

Front: High coils and stabilizer bar
Rear: Leafs

PERFORMANCE MODELS

Sprint—Option included 260/164 V-8; chromed engine trim; bucket-type front seats, console; bucket-styled rear seat cushion; 6000-rpm tachometer; simulated wood steering wheel; Sprint identification on instrument panel, steering wheel hub, front fenders and roof (hardtop); chrome rocker panel trim; wire wheel covers; heavy-duty suspension; heavy rear axle; less restricted air cleaner and muffler; torque boxes in chassis.

EMBLEMS

260 V-8: V-8 symbol over horizontal checked background on front fenders for Sprints. V-8 symbol for others.

OPTIONS

170/101 six, $37.40; 260/164 V-8, $196.30 (conv. $158.90); 4-speed transmission, $90.10 (six), $188.00 (V-8); Fordomatic $163.10 (six), $172.90 (V-8).

1964 FALCON

DIMENSIONS

WB: 109.5"
OL: 181.6, 190 (wagon)
OW: 71.6
FT: 55
RT: 56
HT: 54.5 (sedan), 53.2 (hardtop), 53.8 (convertible), 54.9 (wagon)

MODELS

Series	Body Style	Model No.	Price
Falcon	2-dr sedan*	62A	$1985
Futura	2-dr sedan	62B	2116
Futura	2-dr sedan (Sports)	62C	NA
Futura	2-dr hardtop	63B	2198
Futura	2-dr hardtop (Sports)**	63C	2314
Futura	2-dr convertible	76A	2470
Futura	2-dr convertible (Sports)**	76B	2586

*Convenience Option available ($43.50)
**Sprint Option available ($275—63C; $238—76B)

ENGINES

Code	Type	CID	BxS	Hp@Rpm	Tq@Rpm	CR	Carb
S	I-6	144	3.5x2.5	85@4200	134@2000	8.7	1-bbl
U	I-6	170	3.5x2.94	101@4400	156@2400	9.1	1-bbl
T	I-6	200*	3.68x3.13	116@4400	185@2400	8.7	1-bbl
F	V-8	260	3.8x2.87	164@4400	258@2200	8.8	2-bbl

*Opt, for wagon only

TRANSMISSIONS

3-Speed (Std. 144/85, 170/101)
3-Speed, All-synchro (Std. 260/164)
4-Speed (Opt. all, NA wagon)
Fordomatic (Opt. all)

SUSPENSION

Front: High coils and stabilizer bar
Rear: Leafs

PERFORMANCE MODELS

Sprint—Option included 260/164 V-8; chromed engine trim; bucket-type front seats; console; bucket-styled rear seat cushion; tachometer; simulated wood steering wheel; Sprint identification on instrument panel, steering wheel hub, front fenders and trunk; wire wheel covers; heavy-duty suspension; heavy-duty rear axle; torque boxes in chassis.

EMBLEMS

260 V-8: V-8 symbol with horizontal background on front fenders and trunk (plain V-8) for Sprint. 260 V-8 symbol on front fenders for other models.

OPTIONS

170/101 six, $16.80; 200/116 six, $61.80; 260/164 V-8, $169.40; 4-speed transmission, $90.10 (six), $188.00 (V-8); Fordomatic, $167.40 (six), $177.20 (V-8).

1965 FALCON

DIMENSIONS

WB: 109.5"
OL: 181.6, 190 (wagon)
OW: 71.6
FT: 55 (six), 55.6 (V-8)
RT: 56
HT: (Six) 54.5 (sedan), 53.2 (hardtop), 53.8 (convertible), 54.9 (wagon)
(V-8) 54.9 (sedan), 53.2 (hardtop), 54.1 (convertible), 55.2 (wagon)

MODELS

Series	Body Style	Model No.	Price
Falcon	2-dr sedan*	62A	$2020
Deluxe	2-dr sedan	62B	2144
Deluxe	2-dr hardtop†	63B	2226
Deluxe	2-dr convertible†	76A	2481

*Convenience Option available ($43.50)
†Sprint Package available ($222—63B; $273.60—76A)

ENGINES

Code	Type	CID	BxS	Hp@Rpm	Tq@Rpm	CR	Carb
U	I-6	170	3.5x2.94	105@4400	158@2400	9.1	1-bbl
T	I-6	200	3.68x3.13	120@4400	190@2400	9.2	1-bbl
C	V-8	289	4x2.87	200@4400	282@2400	9.3	2-bbl

TRANSMISSIONS

3-Speed (Std. 170/105, 200/120)
3-Speed, All-synchro (Std. 280/200)
4-Speed (Opt. 289/200)
Cruise-O-Matic (Opt. all)

SUSPENSION

Front: High coils and stabilizer bar
Rear: Leafs

PERFORMANCE MODELS

Sprint—Option included 289/200 V-8; bucket-type front seats; console (convertible only); Sprint identification on console, front fenders

EMBLEMS

289 V-8: V-8 and Sprint symbols combined on Sprints, on front fenders. Smaller 289 over a V-symbol on front fenders of other models.

OPTIONS

200/120 six, $45.10; 289/200 V-8, $153.10; 4-speed transmission, $188.00; Cruise-O-Matic, $172.30 (six), $182.20 (V-8); Limited-slip differential, $38.

1960 FORD

DIMENSIONS

WB: 119"
OL: 213.7
OW: 81.5
FT: 61
RT: 60
HT: 55 (sedans, 4-dr hardtop), 54.5 (Starliner), 55.5 (convertible), 56.5 (wagon)

MODELS

Series	Body Style	Model No.	Price
Fairlane	2-dr sedan (Bus. coupe)	64G	$2170
Fairlane	2-dr sedan	64F	2257
Fairlane 500	2-dr sedan	64A	2334
Galaxie	2-dr sedan	62A	2549
Starliner	2-dr hardtop	63A	2610
Sunliner	2-dr convertible	76B	2860

ENGINES

Code	Type	CID	BxS	Hp@Rpm	Tq@Rpm	CR	Carb
V	I-6	223	3.62x3.6	145@4000	206@2000	8.4	1-bbl
W	V-8	292	3.75x3.3	185@4200	292@2200	8.8	2-bbl
X	V-8	352	4x3.5	235@4400	350@2400	8.9	2-bbl
Y	V-8	352	4x3.5	300@4600	381@2800	9.6	4-bbl
(NA)	V-8	352*	4x3.5	360@6000	380@3400	10.6	4-bbl

*Mid-year engine

TRANSMISSIONS

3-Speed (Std. all)
Overdrive (Opt. all)
Fordomatic (Opt. exc. 352/360)
Cruise-O-Matic (Opt. exc. 223/145, 352/360)

SUSPENSION

Front: Conventional coils and stabilizer bar (except sixes, unless wagon)
Rear: Leafs

OPTIONS

292/185, $113.00; 352/235, $147.80; 352/300, $177.40; 352/360, (NA); Overdrive, $108.40; Fordomatic, $179.80 (six), $189.60 (V-8); Cruise-O-Matic, $211.10; Equa-Lock differential, $38.60.

1961 FORD

DIMENSIONS

WB: 119"
OL: 209.9
OW: 79.9
FT: 61
RT: 60
HT: 55 (sedan, hardtop), 54.5 (Starliner), 55.3 (Sunliner), 56.5 (wagon)

MODELS

Series	Body Style	Model No.	Price
Fairlane	2-dr sedan	64F	$2261
Fairlane 500	2-dr sedan	64A	2376
Galaxie	2-dr sedan	62A	2536
Galaxie	2-dr hardtop	65A	2597
Galaxie	2-dr hardtop (Starliner)	63A	2597
Galaxie	2-dr convertible	76B	2847

ENGINES

Code	Type	CID	BxS	Hp@Rpm	Tq@Rpm	CR	Carb
V	I-6	223	3.62x3.6	135@4000	200@2000	8.4	1-bbl
W	V-8	292	3.75x3.3	175@4200	279@2200	8.8	2-bbl
X	V-8	352	4x3.5	220@4400	336@2400	8.9	2-bbl
Z	V-8	390	4.05x3.78	300@4600	427@2800	9.6	4-bbl
P	V-8	390*	4.05x3.78	330@5000	427@3200	9.6	4-bbl
Q	V-8	390	4.05x3.78	375@6000	427@3400	10.6	4-bbl
Q	V-8	390#	4.05x3.78	401@6000	430@3500	10.6	3x2-bbl

*Police option #Mid-year engine, dealer installed

TRANSMISSIONS

3-Speed (Std. all)
Overdrive (Opt. all)
4-Speed (Dealer-installed option, mid-year)
Fordomatic (Opt. exc. 390's)
Cruise-O-Matic (Opt. exc. 223/135, 390/375, 390/401)

SUSPENSION

Front: Conventional coils and stabilizer bar (except sixes unless wagon)
Rear: Leafs

OPTIONS

292/175, $116.00; 352/220, $148.20; 390/300, $196.70; 390/375, NA, Overdrive, $108.40; Fordomatic, $179.80 (six), $189.60 (V-8); Cruise-O-Matic, $212.30; Equa-Lock differential, $38.60.

1962 FORD

DIMENSIONS

WB: 119"
OL: 209.3
OW: 79.2
FT: 61
RT: 60
HT: 54.8 (sedan, hardtop); 55.1 (convertible); 56.8 (wagon)

MODELS

Series	Body Style	Model No.	Price
Galaxie	2-dr sedan	62B	$2453
Galaxie 500	2-dr sedan	62A	2613
Galaxie 500	2-dr hardtop	65A	2674
Galaxie 500	2-dr convertible	76A	2924
Galaxie 500/XL	2-dr hardtop (V-8)*	65B	NA
Galaxie 500/XL	2-dr convertible (V-8)*	76B	NA

*Mid-year model

ENGINES

Code	Type	CID	BxS	Hp@Rpm	Tq@Rpm	CR	Carb
V	I-6	223	3.62x3.6	138@4000	200@2000	8.4	1-bbl
W	V-8	292	3.75x3.3	170@4200	279@2200	8.8	2-bbl
X	V-8	352	4x3.5	220@4400	336@2400	8.9	2-bbl
Z	V-8	390	4.05x3.78	300@4600	427@2800	9.6	4-bbl
P	V-8	390*	4.05x3.78	330@5000	427@3200	9.6	4-bbl
Q	V-8	390#	4.05x3.78	375@6000	430@3400	10.6	4-bbl
M	V-8	390#	4.05x3.78	401@6000	430@3500	11.1	3x2-bbl
B	V-8	406†	4.13x3.78	385@5800	444@3400	11.4	4-bbl
G	V-8	406†	4.13x3.78	405@5800	448@3500	11.4	3x2-bbl

*Police option #Discontinued mid-year †Mid-year engine

TRANSMISSIONS

3-Speed (Std. exc. 406)
Overdrive (Opt. exc. 406)
4-Speed (Only trans. 406, opt. 352 and 390)
Fordomatic (Opt. exc. 390 and 406)
Cruise-O-Matic (Opt. exc. six, 390/375, 390/401, 406)

SUSPENSION

Front: Conventional coils and stabilizer bar (except sixes, unless wagon)
Rear: Leafs

PERFORMANCE MODELS

Galaxie 500/XL—Models included 292/170 V-8; Cruise-O-Matic; bucket-type front seats; console; bucket-styled rear seat; special door trim panels with lights; floor shift control; bright-metal pedal trim; emblems on glove compartment, roof (hardtop), gas filler cover.

EMBLEMS

390 V-8—Thunderbird emblem with crossed flags and 390 numbers in silver, black, red, white on front fenders, behind wheel.
406 V-8—Thunderbird emblem with crossed flags and 427 numbers in gold, black, white on front fenders behind wheel.

OPTIONS

292/170, $109.00; 352/220, $160.50; 390/300, $246.60; 406/385, $430.80; 406/405, $488.70; Overdrive, $108.40; 4-Speed, $188.00.

1963 FORD

DIMENSIONS

WB: 119"
OL: 209.9
OW: 80
FT: 61
RT: 60
HT: 55.5 (sedan, hardtop), 54.5 (Sports hardtop); 54.6 (convertible), 56.9 (wagon)

MODELS

Series	Body Style	Model No.	Price
300	2-dr sedan	62E	$2324
Galaxie	2-dr sedan	62B	2453
Galaxie 500	2-dr sedan	62A	2613
Galaxie 500	2-dr hardtop	65A	2674
Galaxie 500	2-dr hardtop (Sports)*	63B	2674
Galaxie 500	2-dr convertible	76A	2924
Galaxie 500/XL	2-dr hardtop (V-8)	65B	3268
Galaxie 500/XL	2-dr hardtop (Sports, V-8)*	63C	3268

*Mid-year model

ENGINES

Code	Type	CID	BxS	Hp@Rpm	Tq@Rpm	CR	Carb
V	I-6	223	3.62x3.6	138@4200	203@2200	8.4	1-bbl
F	V-8	260*	3.8x2.87	164@4400	258@2200	8.7	2-bbl
C	V-8	289#	4x2.87	195@4400	282@2400	8.7	2-bbl
X	V-8	352	4x3.5	220@4300	336@2600	8.9	2-bbl
Z	V-8	390	4.05x3.78	300@4600	427@2800	9.6	4-bbl
P	V-8	390†	4.05x3.78	330@5000	427@3200	9.6	4-bbl
B	V-8	406*	4.13x3.78	385@5900	444@3400	10.9	4-bbl
G	V-8	406*	4.13x3.78	405@5800	448@3500	10.9	3x2-bbl
Q	V-8	427#	4.23x3.78	410@5600	476@3400	11.6	4-bbl
R	V-8	427#	4.23x3.78	425@6000	480@3700	11.6	2x4-bbl

*Discontinued mid-year #Mid-year engine †Police option

TRANSMISSIONS

3-Speed, All-synchro (Std. exc. 406, 427)
Overdrive (Opt. all)
4-Speed (Only 406, 427, opt. 352, 390)
Fordomatic (Opt. six, 260/164, 289/195)
Cruise-O-Matic (Opt. exc. six, 406, 427)

SUSPENSION

Front: Conventional coils and stabilizer bar.
Rear: Leafs

PERFORMANCE MODELS

Galaxie 500/XL—Models included 260/164 V-8, later changed to 289/195 V-8; Cruise-O-Matic; bucket-type front seats; console; bucket-styled rear seat; floor shift control; special door trim panels with lights; bright-metal pedal trim; full wheel covers with simulated knock-off spinners; XL emblems on glove compartment, roof (hardtops except Sports), gas filler cover.

EMBLEMS

390 V-8—Thunderbird emblem with crossed flags and 390 numbers in silver, black, red, white on front of front fender.
406 V-8—Thunderbird emblem with crossed flags and 406 numbers in gold, black, white on front of front fenders.
427 V-8—Thunderbird emblem with crossed flags and 427 numbers in gold, black, white on front of front fenders.

OPTIONS

260/164, $109; 289/195, $109; 352/220, $160.50; 390/300, $246.50; 427/410, $514.70; 427/425, $570.60; Overdrive, $108.40; 4-Speed, $188; Fordomatic, $179.80 (six), $189.60 (V-8); Cruise-O-Matic, $212.30; Equa-Lock differential $42.50 (through 390); Dual-Drive differential, $110 (dealer installed); Heavy-duty suspension, $14.60; Conversion for floor shifts for three-speed transmission, $47.95; Heavy-duty radiator, $7.90; Heavy-duty brakes, $9.30; Exhaust cut-outs, $55 (427 only, dealer installed); 7.10x15 nylon tires, $15.80.

1964 FORD

DIMENSIONS

WB: 119"
OL: 209.8
OW: 80
FT: 61
RT: 60
HT: 56.5 (sedan), 55.5 (hardtop, convertible), 57.9 (wagon)

MODELS

Series	Body Style	Model No.	Price
Custom	2-dr sedan	62E	$2350
Custom 500	2-dr sedan	62B	2453
Galaxie 500	2-dr sedan	62A	2613
Galaxie 500	2-dr hardtop	63B	2674
Galaxie 500	2-dr convertible	76A	2936
Galaxie 500/XL	2-dr hardtop (V-8)	63C	3222
Galaxie 500/XL	2-dr convertible	76B	3484

ENGINES

Code	Type	CID	BxS	Hp@Rpm	Tq@Rpm	CR	Carb
V	I-6	223	3.62x3.6	138@4200	203@2200	8.4	1-bbl
C	V-8	289	4x2.87	195@4400	282@2400	8.7	2-bbl
X	V-8	352	4x3.5	250@4400	352@2800	9.3	4-bbl
Z	V-8	390	4.05x3.78	300@4600	427@2800	9.6	4-bbl
P	V-8	390*	4.05x3.78	330@5000	427@3200	10.8	4-bbl
Q	V-8	427	4.23x3.78	410@5600	476@3400	11.5	4-bbl
R	V-8	427	4.23x3.78	425@6000	480@3700	11.5	2x4-bbl

*Police option

TRANSMISSIONS

3-Speed, All-synchro (Std. exc. 427)
Overdrive (Opt. six through 390)
4-Speed (Only trans. for 427, opt. 390)
Cruise-O-Matic (Opt. exc. 427)

SUSPENSION

Front: Conventional coils and stabilizer bar.
Rear: Leafs

PERFORMANCE MODELS

Galaxie 500/XL—Models included 289/195; Cruise-O-Matic; bucket-type front seats; console; bucket-styled rear seat, floor shift control; special door trim panels with lights; bright-metal pedal trim; full wheel covers with simulated knock-off spinners; XL emblems on glove compartment, rear fenders and rear panel under trunk.

EMBLEMS

390 V-8—Thunderbird emblem with crossed flags and 390 numbers in silver, black, white on front fender, near bottom, behind front wheel.
427 V-8—Thunderbird emblem with crossed flags and 427 numbers in gold, black, white on front fender, near bottom, behind front wheel.

OPTIONS

289/195, $109; 352/250, $179.70; 390/300, $246.60; 427/410, $514.60;

427/425, $570.60; Overdrive, $108.40; 4-Speed, $188.00; Cruise-O-Matic, $179.80 (six), $189.60 (289/195), $212.30 (352, 390); Equa-Lock differential, $42.50 (NA 427).

1965 FORD

DIMENSIONS

WB: 119"
OL: 210
OW: 77.4
FT: 62
RT: 62
HT: 55.6 (sedan), 54.7 (hardtop), 54.8 (convertible), 56.7 (wagon)

MODELS

Series	Body Style	Model No.	Price
Custom	2-dr sedan	62E	$2361
Custom 500	2-dr sedan	62B	2464
Galaxie 500	2-dr hardtop	63B	2685
Galaxie 500	2-dr convertible	76A	2950
Galaxie 500/XL	2-dr hardtop (V-8)	63C	3233
Galaxie 500/XL	2-dr convertible (V-8)	76B	3498
Galaxie 500 LTD	2-dr hardtop (V-8)	63F	3233

ENGINES

Code	Type	CID	BxS	Hp@Rpm	Tq@Rpm	CR	Carb
V	I-6	240	4x3.18	150@4000	234@2200	9.2	1-bbl
C	V-8	289	4x2.87	200@4400	282@2400	9.3	2-bbl
X	V-8	352	4x3.5	250@4400	352@2800	9.3	4-bbl
Z	V-8	390	4.05x3.78	300@4600	427@2800	10.1	4-bbl
P	V-8	390*	4.05x3.78	330@5000	427@3200	10.1	4-bbl
Q	V-8	427	4.23x3.78	410@4500	476@3400	11.1	4-bbl
R	V-8	427	4.23x3.78	425@6000	480@3700	11.1	2x4-bbl

*Police option

TRANSMISSIONS

3-Speed, All-synchro (Std. exc. 427)
Overdrive (Opt. six, 289, 390/300)
4-Speed (Only trans. for 427's, opt. 390/300)
Cruise-O-Matic (Opt. exc. 427)

SUSPENSION

Front: Conventional coils and stabilizer bar.
Rear: Coils, three links and track bar

PERFORMANCE MODELS

Galaxie 500/XL—Models included 289/200; Cruise-O-Matic; bucket-type front seats; console; bucket-styled rear seat; floor shift control; bright-metal pedal trim; special door trim with lights; chrome rocker panel; full wheel covers.

EMBLEMS

352, 390 V-8—Vertical silver emblem with checkered background on bottom, Thunderbird symbol near middle and engine size on top against red background.
427 V-8—Vertical silver emblem with checkered background on bottom, Thunderbird symbol near middle and engine size on top against gold background.

OPTIONS

289/200, $109; 352/250, $179.70; 390/300, $246.60; 390/330, $334.30; 427/410, $514.70; 427/425, $570.60; Overdrive, $108.40; 4-Speed, $188; Cruise-O-Matic, $179.80 (six), $189.60 (289), $212.30 (352, 390); Traction-Lock differential, $42.50 (NA 427); Heavy-duty radiator, $10.70; Heavy-duty suspension, $13.30; Heavy-duty brakes, $9.30; Transistorized ignition, $76 (427 only).

1966 FORD

DIMENSIONS

WB: 119"
OL: 210
OW: 79
FT: 62
RT: 62
HT: 55.6 (sedan), 54.7 (hardtop), 54.8 (convertible), 56.7 (wagon)

MODELS

Series	Body Style	Model No.	Price
Custom	2-dr sedan	62E	$2363.59
Custom 500	2-dr sedan	62B	2464.01
Galaxie 500	2-dr hardtop	63B	2666.05
Galaxie 500	2-dr convertible	76A	2913.87
Galaxie 500/XL	2-dr hardtop (V-8)	63C	3208.04
Galaxie 500/XL	2-dr convertible (V-8)	76B	3455.86
Galaxie 500 7 Litre	2-dr hardtop (V-8)	63D	3596.10
Galaxie 500 7 Litre	2-dr convertible (V-8)	76D	3844.75
LTD	2-dr hardtop (V-8)	63F	3178.75

ENGINES

Code	Type	CID	BxS	Hp@Rpm	Tq@Rpm	CR	Carb
V	I-6	240	4x3.18	150@4000	234@2200	9.2	1-bbl
C	V-8	289	4x2.87	200@4400	289@2400	9.3	2-bbl
X	V-8	352	4x3.5	250@4400	352@2800	9.3	4-bbl
H	V-8	390	4.05x3.78	275@4400	405@2600	9.5	2-bbl
Z		390	4.05x3.78	315@4600	427@2800	10.5	4-bbl
W	V-8	427	4.23x3.78	410@5600	476@3400	11.1	4-bbl
R	V-8	427	4.23x3.78	425@6000	480@3700	11.1	2x4-bbl
Q	V-8	428	4.13x3.98	345@4600	462@2800	10.5	4-bbl
P	V-8	428*	4.13x3.98	360@5400	459@3200	10.5	4-bbl

*Police option

TRANSMISSIONS

3-Speed, All-synchro (Std. exc. 427, 428)
Overdrive (Opt. six, 289)
4-Speed (Only trans. for 427, opt. 390/315, 428/345, 428/360)
Cruise-O-Matic (Opt. exc. 427)

SUSPENSION

Front: Conventional coils and stabilizer bar.
Rear: Coils, three links and track bar.

PERFORMANCE MODELS

Galaxie 500/XL—Models included 289/200, Cruise-O-Matic, bucket-type front seats; console; bucket-styled rear seat; floor shift control; bright-metal pedal trim; special door trim with lights; chrome rocker panel, rear fender and wheel lip trim; full wheel covers; XL emblem on trunk.
Galaxie 500 7 Litre—Models included 428/345; Cruise-O-Matic; bucket-type front seats; console; bucket-styled rear seat; floor shift control; bright-metal pedal trim; special door trim with lights; simulated wood steering wheel; striping on body sides; chrome rocker panel, rear fender and wheel lip trim; full wheel covers; 7 Litre emblems on left-hand side of grille, at front of front fenders and on trunk in center.

EMBLEMS

390 V-8—Three slanted vertical parallelograms with checkered pattern on the bottom quarter, an overlay with the numbers 390 and red at the top.
427 V-8—Three slanted vertical parallelograms with checkered pattern on the bottom quarter, an overlay with the numbers 427 and red at the top.
428 V-8—Three slanted vertical parallelograms with checkered pattern on the bottom quarter, an overlay with the numbers 428 and red at the top (not used on 7 Litre).

OPTIONS

289/200, $105.96; 352/250, $164.41; 390/275, $206.76; 390/315, $259.03; 427/425, $1074.01; 428/345, $323.23; Overdrive, $105.33; 4-Speed, $182.69; Cruise-O-Matic, $174.47 (six), $183.99 (289), $214.63 (352/250, 390/275, 390/315, 428/345, 428/360).

1967 FORD

DIMENSIONS

WB: 119"
OL: 213
OW: 79
FT: 62
RT: 62
HT: 55.7 (sedan), 54.7 (hardtop), 54.8 (convertible), 56.9 (wagon)

MODELS

Series	Body Style	Model No.	Price
Custom	2-dr sedan	62E	$2440.78
Custom 500	2-dr sedan	62B	2552.60
Galaxie 500	2-dr hardtop	63B	2754.68
Galaxie 500	2-dr convertible	76A	3033.23
XL	2-dr hardtop (V-8)*	63C	3243.24
XL	2-dr convertible (V-8)*	76B	3492.84
LTD	2-dr hardtop (V-8)	63J	3362.47

*7 Litre Sports Package optional

ENGINES

Code	Type	CID	BxS	Hp@Rpm	Tq@Rpm	CR	Carb
V	I-6	240	4x3.16	150@4000	234@2200	9.2	1-bbl
C	V-8	289	4x2.87	200@4400	282@2400	9.3	2-bbl
H	V-8	390	4.05x3.78	270@4400	403@2600	9.5	2-bbl
Z	V-8	390	4.05x3.78	315@4600	427@2800	10.5	4-bbl
W	V-8	427†	4.23x3.78	410@5600	476@3400	11.1	4-bbl
R	V-8	427†	4.23x3.78	425@6000	480@3700	11.1	2x4-bbl
Q	V-8	428	4.13x3.98	345@4600	462@2800	10.5	4-bbl
P	V-8	428*	4.13x3.98	360@5400	459@3200	10.5	4-bbl

†Limited production option *Police option

TRANSMISSIONS

3-Speed, All-synchro (Std. exc. 427, 428)
Overdrive (Opt. six, 289)
4-Speed (Only transmission for 427, opt. 390/315, 428)
Cruise-O-Matic (Opt. exc. 427)

SUSPENSION

Front: Conventional coils and stabilizer bar
Rear: Coils, three links and track bar

PERFORMANCE MODELS

XL—Models included 289/200; SelectShift Cruise-O-Matic; bucket-type front seats; console; bucket-styled rear seat; floor shift control; special door trim with lights; bright-metal trim on pedals; full wheel covers; XL emblem on grille and trunk.
7 Litre Sports Package—Option included 428/345; Cruise-O-Matic; power front disc brakes, wide-oval tires, heavy-duty suspension; simulated wood steering wheel; XL interior; 7 Litre circular emblem in grille.

EMBLEMS

390 V-8—Slanted parallelogram with checkered bottom quarter, 390 numbers and red top in three divisions.
428 V-8—Same as 1966.
428 V-8—Slanted parallelogram with checkered bottom quarter, 428 numbers and red top in three divisions.

OPTIONS

289/200, $106.72; 390/270, $184.97; 390/315, $264.80; 428/345, $351.49; Overdrive, $116.49; 4-Speed, $184.02; Cruise-O-Matic, $188.18 (six), $197.89 (289), $220.17 (V-8 exc. 427); 7 Litre Sports Package, $515.86; Limited-slip differential, $41.60.

1968 FORD

DIMENSIONS

WB: 119"
OL: 213.3
OW: 78
FT: 62
RT: 62
HT: 55.8 (sedan), 54.8 (4-dr hardtop), 54.6 (formal 2-dr hardtop), 53.9 (fastback 2-dr hardtop), 54.4 (convertible), 57.2 (wagon)

MODELS

Series	Body Style	Model No.	Price
Custom	2-dr sedan	62E	$2639.20
Custom 500	2-dr sedan	62B	2754.17
Galaxie 500	2-dr hardtop	65C	2970.99
Galaxie 500	2-dr hardtop (fastback)	63B	2936.26
Galaxie 500	2-dr convertible	76A	3164.73
XL	2-dr hardtop (fastback)*	63C	3040.65
XL	2-dr convertible*	76B	3270.18
LTD	2-dr hardtop (V-8)	65A	3208.30

*GT Equipment Group optional (Prices as of 2/68)

ENGINES

Code	Type	CID	BxS	Hp@Rpm	Tq@Rpm	CR	Carb
V	I-6	240	4x3.16	150@4000	234@2200	9.2	1-bbl
F	V-8	302	4x3	210@4600	300@2600	9	2-bbl
Y	V-8	390	4.05x3.78	265@4400	390@2600	9.5	2-bbl
Z	V-8	390	4.05x3.78	315@4600	427@2800	10.5	4-bbl
W	V-8	427*	4.23x3.78	390@5600	460@3200	10.9	4-bbl
Q	V-8	428	4.13x3.98	340@4600	462@2800	10.5	4-bbl
P	V-8	428#	4.13x3.98	360@5400	459@3200	10.5	4-bbl

*Discontinued #Police option

TRANSMISSIONS

3-Speed, All-synchro (Std. exc. 427, 428)
4-Speed (Opt. 390/315, 428/340)
Cruise-O-Matic (Opt. all)

SUSPENSION

Front: Conventional coils and stabilizer bar
Rear: Coils, three links and track bar

PERFORMANCE MODELS

XL—Models included bucket-type seats and console (changed to an option mid-year); disappearing headlights; XL emblems on hood, trunk and roof (hardtop); full wheel covers.
GT Equipment Group—Option included heavy-duty suspension, power front disc brakes; low-restriction exhaust; wide-oval tires; 3.25:1 axle ratio; simulated mag wheel covers; GT stripes on side of car; GT emblems on front fenders, behind wheel. (Available only with 390, 427 or 428.)

EMBLEMS

390 V-8—Horizontal rectangle with three sections and 390 numbers in each, red background, on front fender, behind wheel.
428 V-8—Horizontal rectangle with three sections and 428 numbers in each, red background, on front fender, behind wheel.

OPTIONS

302/210, $106.72; 390/265, $184.97; 390/315, $264.80; 428/340, $351.49; Heavy-duty 3-speed, $79.20; 4-Speed, $184.02; Cruise-O-Matic, $191.13 (six); $200.85 (302/210), $223.03 (390/265), $233.17 (390/315, 427, 428); GT Equipment Group, $204.64; Limited-slip differential, $41.60.

1969 FORD

DIMENSIONS

WB: 121"
OL: 213.9 (Custom, Galaxie 500), 216.9 (XL, LTD, wagon exc. Squire), 219 (Squire)
OW: 79.4 (2-dr sedan, convertible), 79.7 (2-dr hardtop), 79.8 (all 4-door)
FT: 63
RT: 64
HT: 54.6 (2-dr sedan), 54.9 (4-dr sedan), 53.7 (2-dr hardtop), 53.8 (4-dr hardtop); 54 (convertible), 56.8 (wagon)

MODELS

Series	Body Style	Model No.	Price
Custom	2-dr sedan	62E	$2649
Custom 500	2-dr sedan	62B	2748
Galaxie 500	2-dr hardtop	65C	2982
Galaxie 500	2-dr hardtop (SportsRoof)	63B	2930
Galaxie 500	2-dr convertible	76A	3159
XL	2-dr hardtop (SportsRoof)*	63C	3069
XL	2-dr convertible*	76B	3279
LTD	2-dr hardtop (V-8)	65A	3152

*GT Performance Group optional (Prices as of 2/69)

ENGINES

Code	Type	CID	BxS	Hp@Rpm	Tq@Rpm	CR	Carb
V	I-6	240	4x3.18	150@4000	234@2200	9.2	1-bbl
F	V-8	302	4x3	220@4600	300@2600	9.5	2-bbl
H	V-8	351*	4x3.5	250@4600	355@2600	9.5	2-bbl
Y	V-8	390	4.13x3.78	265@4400	390@2600	9.5	2-bbl
P	V-8	428#	4.13x3.98	360@5400	459@3200	10.5	4-bbl
K	V-8	429	4.36x3.59	320@4400	460@2200	10.5	2-bbl
N	V-8	429	4.36x3.59	360@4600	480@2800	10.5	4-bbl

*Mid-year engine #Police option

TRANSMISSIONS

3-Speed, all-synchro (Std. six, 302, 390)
4-Speed (Opt. 429/360)
Cruise-O-Matic (Opt. all)

SUSPENSION

Front: Conventional coils and stabilizer bar
Rear: Coils, three links and track bar

PERFORMANCE MODELS

XL—Models included disappearing headlights; full wheel covers; XL emblems on doors, hood, trunk.
GT Performance Group—Option included Competition suspension; power front disc brakes; H70x15 belted tires; GT emblems on side trim. (Available only with 390 or 429)

EMBLEMS

429 V-8—On black background trim piece next to front side-marker light with 429 numbers listed.

OPTIONS

302/220, $105; 351/250, $158.18; 390/265, $163.34; 429/320, $268.24; 429/360, $342.07; 4-Speed, $194.31; Cruise-O-Matic, $191.13 (six), $200.85 (302, 351, 390), $222.08 (428, 429); GT Performance Group, $259.09; Bucket seats and console, $168.62 (Galaxie 500 and XL 2-dr); Limited-slip differential, $41.60; Power front disc brakes, $64.77.

1970 FORD

DIMENSIONS

WB: 121"
OL: 213.9 (Custom, Galaxie 500), 216.9 (XL, LTD, wagon exc. Squire), 219 (Squire)
OW: 79.7 (2-dr), 79.8 (4-dr exc. Squire), 80 (Squire)
FT: 63
RT: 64
HT: 54.9 (4-dr sedan), 53.5 (hardtop), 53.9 (convertible), 56.9 (wagon)

MODELS

Series	Body Style	Model No.	Price
Galaxie 500	2-dr hardtop	65C	$3094
Galaxie 500	2-dr hardtop (SportsRoof)	63B	3043
XL	2-dr hardtop (Sports-Roof, V-8)	63C	3293
XL	2-dr convertible (V-8)	76B	3501
LTD	2-dr hardtop (V-8)	65A	3356

(Prices as of 12/19/69)

ENGINES

Code	Type	CID	BxS	Hp@Rpm	Tq@Rpm	CR	Carb
V	I-6	240	4x3.18	150@4000	234@2200	9.2	1-bbl
F	V-8	302	4x3	220@4600	300@2600	9.5	2-bbl
H	V-8	351	4x3.5	250@4600	355@2600	9.5	2-bbl
Y	V-8	390	4.05x3.78	265@4400	390@2600	9.5	2-bbl
P	V-8	428*	4.13x3.98	360@5400	459@3200	10.5	4-bbl
K	V-8	429	4.36x3.59	320@4400	460@2800	10.5	2-bbl
N	V-8	429	4.36x3.59	360@4600	480@2800	10.5	4-bbl

*Police option

TRANSMISSIONS

3-Speed, All synchro (Std. exc. 428, 429)
Cruise-O-Matic (Opt. all)

SUSPENSION

Front: Conventional coils and stabilizer bar
Rear: Coils, three links and track bar

PERFORMANCE MODELS

XL—Models included 351/250; disappearing headlights; full wheel covers, XL logos on hood, trunk and roof (hardtop).

OPTIONS

302/220, $79 (Custom, Custom 500 sedan only); 351/250, $124 (Custom), $111 (Galaxie 500); 390/265, $210 (Custom), $197 (Galaxie 500); 429/320; $292 (Custom), $279 (Galaxie 500); 429/360, $366 (Custom), $353 (Galaxie 500); Cruise-O-Matic, $201 (exc. 428, 429), $222 (429); Bucket seats and console (Galaxie 500 and XL 2-dr only), $188; Heavy-duty suspension, $23; Traction-Lok differential, $43; Dualtone paint, $45 (XL only). (Prices as of 12/19/69)

1962 FAIRLANE

DIMENSIONS

WB: 115.5"
OL: 197.6
OW: 71.3
FT: 57
RT: 56
HT: 55.4

MODELS

Series	Body Style	Model No.	Price
Fairlane	2-dr sedan	62A	$2154
Fairlane 500	2-dr sedan	62B	2242
Fairlane 500	2-dr sedan (Sports Coupe)*	62C	2403

*Mid-year model

ENGINES

Code	Type	CID	BxS	Hp@Rpm	Tq@Rpm	CR	Carb
U	I-6	170	3.5x2.94	101@4400	156@2400	8.7	1-bbl
L	V-8	221	3.5x2.87	145@4400	216@2200	8.7	2-bbl
F	V-8	260*	3.8x2.87	164@4400	258@2200	8.7	2-bbl

*Mid-year engine

TRANSMISSIONS

3-Speed (Standard)
Overdrive (Opt. 221/145 only)

Fordomatic (Opt. all)

SUSPENSION

Front: High coils and stabilizer bar
Rear: Leafs

PERFORMANCE MODELS

Fairlane 500 Sports Coupe—Model included bucket-style front seats; console between the seats; Sports Coupe nameplate on right side of trunk.

EMBLEMS

221 V-8—Small V-8 symbol on front fender behind wheel (early models did not have any emblem for a V-8).
260 V-8—V-8 symbol with overlay of numbers 260 in red on front fender behind wheel.

OPTIONS

221/145, $103; 260/164, $140.40; Overdrive, $108.40; Fordomatic, $179.80 (six), $189.60 (V-8).

1963 FAIRLANE

DIMENSIONS

WB: 115.5"
OL: 197.6 (sedan and hardtop), 201.8 (wagon)
OW: 71.3
FT: 57
RT: 56
HT: 55.4 (sedan), 54.5 (hardtop), 56.6 (wagon)

MODELS

Series	Body Style	Model No.	Price
Fairlane	2-dr sedan	62A	$2154
Fairlane 500	2-dr sedan	62B	2242
Fairlane 500	2-dr hardtop	65A	2324
Fairlane 500	2-dr hardtop (Sports Coupe)	65B	2504

ENGINES

Code	Type	CID	BxS	Hp@Rpm	Tq@Rpm	CR	Carb
U	I-6	170	3.5x2.94	101@4400	156@2400	8.7	1-bbl
T	I-6	200*	3.68x3.13	116@4400	185@2400	8.7	1-bbl
L	V-8	221	3.5x2.87	145@4400	216@2000	8.7	2-bbl
F	V-8	260	3.8x2.87	164@4400	258@2200	8.7	2-bbl
K	V-8	289*	4x2.87	271@6000	312@3400	11.6	4-bbl

*Mid-year engine

TRANSMISSIONS

3-Speed (Std. 170/101)
3-Speed, All-synchro (Std. all V-8's)
Overdrive (Opt. any V-8)
Fordomatic (Opt. exc. six in wagon, 289; only trans. for 200/116)

SUSPENSION

Front: High coils and stabilizer bar
Rear: Leafs

PERFORMANCE MODELS

Fairlane 500 Sports Coupe—Model included front bucket-type seats; bucket-styled rear seat; console from between front seats to under dash; deluxe wheel covers with simulated knock-off spinners; three "portholes" in side trim on front fender; red, white and blue emblem on roof; Sports Coupe nameplate on right side of trunk.

EMBLEMS

260 V-8—V-8 symbol with 260 numbers overlaid in red on front of front fender.
289 V-8—V-8 symbol with the numbers 289 on front of front fender.

OPTIONS

200/116, (NA); 221/145, $103; 260/164, $154.50; 289/271, $424.80; Overdrive, $108.40 (V-8 only); 4-Speed $188 (V-8 only); Fordomatic, $179.80 (six), $189.60 (V-8 exc. 289); Heavy-duty suspension, $14.60; Heavy-duty radiator, $5.70; 7.00x14 tires and wheels, $7.40.

1964 FAIRLANE

DIMENSIONS

WB: 115.5"
OL: 197.6 (sedan and hardtop), 201.8 (wagon)
OW: 72.2
FT: 57
RT: 56
HT: 55.8 (sedan), 54.9 (hardtop), 56.6 (wagon)

MODELS

Series	Body Style	Model No.	Price
Fairlane	2-dr sedan	62A	$2183
Fairlane 500	2-dr sedan	62B	2265
Fairlane 500	2-dr hardtop	65A	2330
Fairlane 500	2-dr hardtop (Sports Coupe)	65B	2491

ENGINES

Code	Type	CID	BxS	Hp@Rpm	Tq@Rpm	CR	Carb
U	I-6	170	3.5x2.94	101@4400	156@2400	8.7	1-bbl
T	I-6	200	3.68x3.13	116@4400	185@2400	8.7	1-bbl
F	V-8	260	3.8x2.87	164@4400	258@2200	8.8	2-bbl
C	V-8	289	4x2.87	195@6000	282@2400	9.0	2-bbl
K	V-8	289	4x2.87	271@6000	312@3400	10.5	4-bbl

TRANSMISSIONS

3-Speed (Std. 170/101)
3-Speed, All-synchro (Std. all V-8)
Overdrive (Opt. 260/164)
4-Speed (Opt. 289)
Fordomatic (Only trans. for 200/116, opt. 260/164)
Cruise-O-Matic (Opt. 289/195)

SUSPENSION

Front: High coils and stabilizer bar
Rear: Leafs

PERFORMANCE MODELS

Fairlane 500 Sports Coupe—Model included bucket-type front seats; console; bucket-styled rear seat; full wheel covers with spinners; lights in interior door panels; Sports Coupe nameplate and red, white and blue emblem on roof pillar; Sports Coupe plaque below trunk, next to right taillight.

EMBLEMS

260 V-8—V-symbol with 260 numbers in center, on front of front fender.
289 V-8—V-symbol with 289 numbers in center, on front of front fender.

OPTIONS

260/164, $100; 289/195, $145.10; 289/271, $421.80; Overdrive, $108.40; 4-Speed, $188; Fordomatic, $189.60; Cruise-O-Matic, $189.60.

1965 FAIRLANE

DIMENSIONS

WB: 116"
OL: 198.8 (sedan and hardtop), 203 (wagon)
OW: 73.8
FT: 57
RT: 57
HT: 55.7 (sedan), 54.8 (hardtop), 56.7 (wagon)

MODELS

Series	Body Style	Model No.	Price
Fairlane	2-dr sedan	62A	$2230
Fairlane 500	2-dr sedan	62B	2312
Fairlane 500	2-dr hardtop	65A	2377
Fairlane 500	2-dr hardtop (Sports Coupe)	65B	2538

ENGINES

Code	Type	CID	BxS	Hp@Rpm	Tq@Rpm	CR	Carb
T	I-6	200	3.68x3.13	120@4400	190@2400	9.2	1-bbl
C	V-8	289	4x2.87	200@4400	282@2400	9.3	2-bbl
A	V-8	289	4x2.87	225@4800	305@3200	10.0	4-bbl
K	V-8	289	4x2.87	271@6000	312@3400	10.5	4-bbl

TRANSMISSIONS

3-Speed (Std. six)
3-Speed, All-synchro (Std. 289/200, 289/225)
Overdrive (Opt. 289/200)
4-Speed (Opt. 289/225, only trans. for 289/271)
Cruise-O-Matic (Opt. all)

SUSPENSION

Front: High coils and stabilizer bar
Rear: Leafs

PERFORMANCE MODELS

Fairlane 500 Sports Coupe—Model included bucket-type front seats; console; bucket-styled rear seat; wheel covers; lights in interior door panels; Sports Coupe nameplate and red, white and blue emblem on roof pillar; Sports Coupe nameplate on right side of trunk; three fluted additions to side trim on front fender.

EMBLEMS

289 V-8—V-symbol with twin checkered flags below each side and the numbers 289 above it, on front fender, behind wheel.

OPTIONS

289/200, $108; 289/225, $162; 289/271, $429.80; Overdrive, $108.40; 4-Speed, $188; Cruise-O-Matic, $179.80 (six), $189.60 (V-8); Limited-slip differential, $38 (NA 289/271); Heavy-duty suspension, $13.30; Heavy-duty brakes, $9.30; Heavy-duty radiator, $10.70.

1966 FAIRLANE

DIMENSIONS

WB: 116", 113 (wagon)
OL: 197, 199.8 (wagon)
OW: 74 (2-dr), 74.1 (GT and GTA), 74.7 (4-dr)
FT: 58
RT: 58
HT: 55 (sedan), 54.3 (hardtop), 54 (convertible), 55.9 (wagon)

MODELS

Series	Body Style	Model No.	Price
Fairlane	2-dr sedan	62A	$2223.69
Fairlane 500	2-dr sedan	62B	2300.19
Fairlane 500	2-dr hardtop	63B	2361.24
Fairlane 500	2-dr convertible	76B	2584.36
Fairlane 500/XL	2-dr hardtop	63C	2524.80
Fairlane 500/XL	2-dr convertible	76C	2747.92
Fairlane GT*	2-dr hardtop (V-8)	63D	2822.65
Fairlane GT*	2-dr convertible (V-8)	76D	3045.76

*Becomes GTA with automatic trans.

ENGINES

Code	Type	CID	BxS	Hp@Rpm	Tq@Rpm	CR	Carb
T	I-6	200	3.68x3.13	120@4400	190@2400	9.2	1-bbl
C	V-8	289	4x2.87	200@4400	282@2400	9.3	2-bbl
Y	V-8	390*	4.05x3.78	265@4400	401@2600	9.5	2-bbl
Y	V-8	390*	4.05x3.78	275@4400	405@2600	9.5	2-bbl
Z	V-8	390	4.05x3.78	315@4600	427@2800	10.5	4-bbl
S	V-8	390	4.05x3.78	335@4800	427@3200	10.5	4-bbl
W	V-8	427#	4.23x3.78	410@5600	476@3400	11.1	4-bbl
R	V-8	427#	4.23x3.78	425@6000	480@3700	11.1	2x4-bbl

*265 hp with stick, 275 hp with automatic
#Mid-year engine, limited production

TRANSMISSIONS

3-Speed (Std. six)
3-Speed, All-synchro (Std. V-8 exc. 427)
Overdrive (Opt. 289/200)
4-Speed (Opt. V-8 only)
Cruise-O-Matic (Opt. 200/120, 289/200, 390/275, 390/315)
Sport Shift Cruise-O-Matic (Opt. 390/335)

SUSPENSION

Front: High coils and stabilizer bar
Rear: Leafs

PERFORMANCE MODELS

Fairlane 500/XL—Models included front bucket-type seats; console; special interior door panels with red and white lights; round XL emblem in center of grille; XL nameplate on rear fender; XL badge on right side of trunk; deluxe full wheel covers.
Fairlane GT/GTA—Models included 390/335; chromed engine trim; simulated hood-top louvers with engine-size numbers; bucket-type front seats; console; special interior door trim with red and white lights; Deluxe full wheel covers; blacked-out grille with GT badge; side accent stripes with GT or GTA badges on front fender; GT or GTA badge on right side of trunk.

EMBLEMS

289 V-8—V-symbol with twin checkered flags below each side and numbers 289 above small gold inverted triangle, on front fender, behind wheel.
390 V-8—V-symbol with twin checkered flags below each side and numbers 390 above small red inverted triangle, on front fender, behind wheel. (Not used on GT/GTA.)
427 V-8—Emblem from full-sized Ford 427 on front of front fender.

OPTIONS

289/200, $104.84; 390/265, $205.64; 390/315, $257.91; Overdrive, $105.23; 4-Speed, $182.69; Cruise-O-Matic, $174.47 (six), $183.99 (289/200), $214.63 (390); Limited-slip differential, $36.93; Styled steel wheels, $93.16 (XL and GT), $114.28 (others).

1967 FAIRLANE

DIMENSIONS

WB: 116", 113 (wagon)
OL: 197, 199.8 (wagon)
OW: 74 (2-dr), 74.7 (4-dr)
FT: 58.4, 58.5 (GT)
RT: 58.1, 58.6 (GT)
HT: 55.2 (sedan), 54.2 (hardtop), 54.6 (convertible), 56.2 (wagon)

MODELS

Series	Body Style	Model No.	Price
Fairlane	2-dr sedan	62A	$2296.52
Fairlane 500	2-dr sedan	62B	2376.63
Fairlane 500	2-dr hardtop	63B	2439.14
Fairlane 500	2-dr convertible	96B	2663.94
Fairlane 500/XL	2-dr hardtop	63C	2618.52
Fairlane 500/XL	2-dr convertible	76C	2843.32
Fairlane GT*	2-dr hardtop (V-8)	63D	2838.88
Fairlane GT*	2-dr convertible (V-8)	76D	3063.67

*Becomes GTA with automatic transmission

ENGINES

Code	Type	CID	BxS	Hp@Rpm	Tq@Rpm	CR	Carb
T	I-6	200	3.68x3.13	120@4400	190@2400	9.2	1-bbl
C	V-8	289	4x2.87	200@4400	282@2400	9.3	2-bbl
Y, H	V-8	390	4.05x3.78	270@4400	403@2600	9.5	2-bbl
S	V-8	390	4.05x3.78	320@4800	427@3200	10.5	4-bbl
W	V-8	427*	4.23x3.78	410@5600	476@3400	11.1	4-bbl
R	V-8	427*	4.23x3.78	425@6000	480@3700	11.1	2x4-bbl

*Limited production engine

TRANSMISSIONS

3-Speed, All-synchro (Std. exc. 427)
Overdrive (Opt. 289/200)
4-Speed (Opt. all V-8, only trans. for 427)
Cruise-O-Matic (Opt. exc. 427)

SUSPENSION

Front: High coils and stabilizer bar
Rear: Leafs

PERFORMANCE MODELS

Fairlane 500/XL—Models included front bucket-type seats; console; special interior door panels with red and white lights; XL badges in center of grille, on rear fenders and on trunk, next to right taillight; wheel covers.
Fairlane GT/GTA—Models included 289/200; power front disc brakes; wide-oval tires; front bucket-type seats; special interior door panels with red and white lights; power dome hood with built-in directional signal indicators and engine size when equipped with a 390; blacked-out grille; GT badge in center of grille and GT or GTA badge on trunk next to right taillight; side accent stripes with GT or GTA badges on front fender; Deluxe full wheel covers.

EMBLEMS

289 V-8—Crest with numbers 289 over top part, blue background, on front of front fender except GT, GTA.
390 V-8—Crest with numbers 390 over top part, red background, on front of front fender, except GT, GTA.
427 V-8—Emblem from 1966 full-sized Ford on front of front fender.

OPTIONS

289/200, $106.72; 390/270, $183.88; 390/320, $263.71; Overdrive, $116.59; 4-Speed, $184.02; Cruise-O-Matic, $188.18, $197.89 (289/200), $220.17 (390/270, 390/320); Heavy-duty 3-Speed, $79.20 (required 390/320); Power disc front brakes, $97.21; Console, $47.92 (GT only); Styled steel wheels, $93.84 (XL), $74.36 (GT), $115.11 (others); Wide-oval tires, $83.12.

1968 FAIRLANE

DIMENSIONS

WB: 116", 113 (wagon)
OL: 201.1, 203.9 (wagon)
OW: 74.6 (2-dr), 74.8 (4-dr)
FT: 58.8
RT: 58.9
HT: 55 (sedan), 53.5 (hardtop), 54.5 (convertible), 56 (wagon)

MODELS

Series	Body Style	Model No.	Price
Fairlane	2-dr hardtop	65A	$2502.22
Fairlane 500	2-dr hardtop	65B	2637.07
Fairlane 500	2-dr hardtop (Fastback)	63B	2611.74
Fairlane 500	2-dr convertible	76B	2869.49
Torino	2-dr hardtop	65C	2756.32
Torino GT	2-dr hardtop (V-8)	65D	2818.37
Torino GT	2-dr hardtop (fastback, V-8)	63D	2793.04
Torino GT	2-dr convertible (V-8)	76D	3048.57

(Prices as of 2/68)

ENGINES

Code	Type	CID	BxS	Hp@Rpm	Tq@Rpm	CR	Carb
T	I-6	200	3.68x3.13	115@3800	190@2200	8.8	1-bbl
C	V-8	289*	4x2.87	195@4600	288@2600	8.7	2-bbl
F	V-8	302	4x3	210@4600	300@2600	9.0	2-bbl
Y, H	V-8	390	4.05x3.78	265@4400	390@2600	9.5	2-bbl
S	V-8	390	4.05x3.78	325@4800	427@3200	10.5	4-bbl
W	V-8	427#	4.23x3.78	390@5600	460@3200	10.9	4-bbl
Q	V-8	428*	4.13x3.98	335@5400	440@3400	10.6	4-bbl

*Mid-year engine #Discontinued mid-year

TRANSMISSIONS

3-Speed, All-synchro (Std. exc. 427, 428)
4-Speed (Opt. exc. 427)
Cruise-O-Matic (Opt. all, only trans. for 427)

SUSPENSION

Front: High coils and stabilizer bar
Rear: Leafs

PERFORMANCE MODELS

Torino GT—Models included 302/210 (changed mid-year to 289/195); bucket-type front seats and console (changed mid-year to an option); styled steel wheels with GT emblem on hubcaps; wide-oval tires; C-stripes on sides (fastback); rocker panel and rear fender stripes (hardtop and convertible, available early in model year for fastback); Grand Touring emblems on inner door panels; GT badge in grille; GT round badge in center between taillights (fastback); GT emblem in center of trunk (hardtop and convertible); round emblem on roof pillar (hardtop).

EMBLEMS

302 V-8—Crest with numbers 302 over top part, mounted on front of front fender (discontinued mid-year).
390 V-8—Crest with numbers 390 over top part, mounted on front of front fender.

OPTIONS

289/195, $87.59; 302/210, $113.50; 390/265, $184.59; 390/325, $264.42; 4-Speed, $184.02; Cruise-O-Matic, $191.13 (six), $200.85 (289/195, 302/210), $223.03 (390/265), $233.17 (390/325); Heavy-duty 3-Speed, $79.20 (required 390/325); Power front disc brakes, $64.77; Bucket-type seats, $110.16 (Fairlane 500, GT 2-door). (Prices as of 2/68)

1969 FAIRLANE

DIMENSIONS

WB: 116", 113 (wagon)
OL: 201.1 (sedan, hardtop, convertible), 203.9 (wagon), 206 (Talladega)
OW: 74.6 (2-door), 74.8 (4-door)
FT: 58.8
RT: 58.5
HT: 53.7 (4-dr sedan), 52.4 (hardtop), 52.2 (fastback), 52.8 (convertible), 56.1 (wagon), 52.6 (Talladega)

MODELS

Series	Body Style	Model No.	Price
Fairlane	2-dr hardtop	65A	$2499
Fairlane 500	2-dr hardtop	65B	2626
Fairlane 500	2-dr hardtop (SportsRoof)	63B	2601
Fairlane 500	2-dr convertible	76B	2851
Torino	2-dr hardtop	65C	2754
Torino GT	2-dr hardtop (V-8)	65D	2865
Torino GT	2-dr hardtop (Sports-Roof, V-8)	63D	2840
Torino GT	2-dr convertible (V-8)	76D	3090

Cobra	2-dr hardtop (V-8)	*	3206
Cobra	2-dr hardtop (SportsRoof, V-8)	**	3181
Talladega	2-dr hardtop (V-8)	***	NA

*Regular production option for Fairlane 500 2-dr hardtop (65B)
**Regular production option for Fairlane 500 SportsRoof (63B)
***Domestic Special Order option for Fairlane 500 SportsRoof (63B)
(Prices as of 2/69)

ENGINES

Code	Type	CID	BxS	Hp@Rpm	Tq@Rpm	CR	Carb
L	I-6	250	3.68x3.91	155@4400	240@1600	9.0	1-bbl
F	V-8	302	4x3	220@4600	300@2600	9.5	2-bbl
H	V-8	351	4x3.5	250@4600	355@2600	9.5	2-bbl
M	V-8	351	4x3.5	290@4800	385@3200	10.7	4-bbl
S	V-8	390	4.05x3.78	320@4600	427@3200	10.5	4-bbl
Q	V-8	428	4.13x3.98	335@5200	440@3400	10.6	4-bbl
R	V-8	428*	4.13x3.98	335@5200	440@3400	10.6	4-bbl

TRANSMISSIONS

3-Speed, All-synchro (Std. exc. 428)
4-Speed (Opt. 351, 390, 428)
Cruise-O-Matic (Opt. all)

SUSPENSION

Front: High coils and stabilizer bar
Rear: Leafs

PERFORMANCE MODELS

Torino GT—Models included 302/220; styled steel wheels with GT emblem on hubcaps; F70x14 belted whitewall tires; C-stripes on SportsRoof; rocker panel and rear fender stripes on hardtop and convertible; Grand Touring emblems on inner door panels; GT letters on right side of grille; GT round badge in center between taillights (SportsRoof); GT round badge on roof pillars (hardtop); GT emblem in center of trunk (hardtop and convertible); non-functional hood scoop with turn signal indicators on back; GT suspension.

Torino Cobra—Option included 428/335; 4-Speed; 6-inch wheel rims; F70x14 belted whitewall tires; competition suspension; hood lock pins; blacked-out grille; Cobra emblem on front fenders, behind wheels and on rear panel, next to right taillight (early models had decals).

Torino Talladega—Option included 428/335; Cruise-O-Matic; styled steel wheels with plain hubcaps; F70x14 belted whitewall tires; power front disc brakes; competition suspension; extended nose; blacked-out grille; flat-black hood; flow-stripe at beltline; T-emblem above exterior door handles; round T-badge on panel between taillights; Talladega identification plate on interior door panels; power steering; engine oil cooler; cloth and vinyl bench seats.

EMBLEMS

351 V-8—Crest with numbers 351 over top part, on front of front fenders.

390 V-8—Crest with numbers 390 over top part, on front of front fenders.

428 V-8—Numbers 428 in box when car had scoop, Cobra Jet lettering on both sides of scoop. When no scoop, numbers 428 in small chrome box with red background and individual box for each letter, on front of front fender. (Some early 428's without scoops did not have emblems).

OPTIONS

302/220, $90; 351/250, $144.34; 351/290, $174.25; 390/320, $253.24; 428/335, $377.53; 428/335 Ram-Air, $510.96; 4-Speed, $194.31; Cruise-O-Matic, $191.13 (six), $200.85 (302, 351), $222.08 (390, 428); Heavy-duty 3-Speed, $79.20 (required 390/320); Power front disc brakes, $64.77; Bucket-type seats, $120.59 (Fairlane 500, GT, Cobra); Console, $53.82 (with bucket option only); Tachometer, $47.92; Traction-Lok differential, $63.51 (351, 390, 428); Limited-slip differential, $41.60 (six, 302); Styled steel wheels argent, $38.86; Chrome or color-keyed styled steel wheels, $116.59; Competition suspension, $30.64; GT Handling suspension, $30.64. (Prices as of 2/69)

1970 TORINO

DIMENSIONS

WB: 117", 114 (wagon)
OL: 206.2, 209 (wagon)
OW: 76.4 (4-dr sedan, hardtop), 76.7 (2-dr hardtop, convertible), 76.8 (SportsRoof), 75.4 (wagon)
FT: 60.5
RT: 60.0
HT: 53 (4-dr sedan, hardtop), 52.2 (2-dr hardtop), 51.1 (SportsRoof), 52.8 (convertible), 55.7 (wagon)

MODELS

Series	Body Style	Model No.	Price
Falcon (1970½)*	2-dr sedan	62A	$2460
Fairlane 500	2-dr hardtop	65B	2660
Torino	2-dr hardtop	65C	2722
Torino	2-dr hardtop (SportsRoof)†	63C	2810
Torino Brougham	2-dr hardtop (V-8)	65E	3006
Torino GT	2-dr hardtop (SportsRoof, V-8)	63F	3105
Torino GT	2-dr convertible (V-8)	76F	3212
Torino Cobra	2-dr hardtop (SportsRoof, V-8)	63H	3270

*Mid-year model incorporated into Torino line late in model year
†Mid-year model (Prices as of 12/19/69)

ENGINES

Code	Type	CID	BxS	Hp@Rpm	Tq@Rpm	CR	Carb
L	I-6	250	3.68x3.91	155@4000	240@1600	9.0	1-bbl
F	V-8	302	4x3	220@4600	300@2600	9.5	2-bbl
H	V-8	351	4x3.5	250@4600	355@2600	9.5	2-bbl
M	V-8	351	4x3.5	300@5400	380@3400	11.0	4-bbl
N	V-8	429	4.36x3.59	360@4600	480@2800	10.5	4-bbl
C	V-8	429	4.36x3.59	370@5400	450@3400	11.3	4-bbl
J	V-8	429*	4.36x3.59	370@5400	450@3400	11.3	4-bbl
C**	V-8	429†	4.36x3.59	375@5600	450@4000	11.3	4-bbl
J**	V-8	429†*	4.36x3.59	375@5600	450@4000	11.3	4-bbl

*Ram-Air †Mid-year engine **Axle code has V(3.91) or Y (4.11)

TRANSMISSIONS

3-Speed, All-synchro (Std. exc. 429)
4-Speed (Opt. exc. six, 302)
Cruise-O-Matic (Opt. all)

SUSPENSION

Front: High coils and stabilizer bar
Rear: Leafs

PERFORMANCE MODELS

Torino GT—Models included 302/220; GT hood with integral non-functional scoop; round GT badge in grille; GT letters on sides, near front of rear quarter panel; wheel lip and rocker panel chrome moldings; full-width taillight effect with honeycomb applique and reflective center insert between taillights; black deck lid appliques (SportsRoof); color-keyed racing mirrors; hubcaps and wheel trim rings.

Torino Cobra—Model Included 429/360; Quick-ratio 4-speed; Hurst shifter; 7-inch wide wheels, hubcaps; F70x14 raised-lettering tires; Competition suspension; flat-black hood and grille; twist-type exposed hood latches; wheel lip moldings; Cobra decals on front fenders behind wheels and on panel next to right taillight.

EMBLEMS

351 V-8—Small chrome rectangle on lower front fender, with numbers 351.

429 V-8—Small chrome rectangle on lower front fenders with numbers 429.

When optional shaker hood scoop is used with a Cobra Jet engine (429/370, 429/375) both sides of the scoop have Cobra Jet nameplate.

OPTIONS

302/220, $90; 351/250, $135; 351/300, $183; 429/360, $282; 429/370, $446; 429/370 Ram-Air, $501; 4-Speed with Hurst shifter, $194; Cruise-O-Matic, $201 (six, 302, 351); $222 (429); Power front disc brakes, $65; Drag Pack differential, 3.91:1 Traction-Lok, $155; Drag Pack differential, 4.30:1 Detroit Locker, $207; Heavy-duty suspension, $23 (NA 429); Shaker hood scoop, $65 (351/300); Tachometer, 8000 rpm, $49; Styled steel wheels, argent, $58; Magnum 500 chrome wheels, $155; Bucket-type front seats, $133 (2-dr hardtop except Fairlane 500, SportsRoof and convertible); Console, $54 (same models as buckets); Laser Stripe $39 (GT only); Sports Slats, $65 (SportsRoof). (Prices as of 12/19/69)

1971 TORINO

DIMENSIONS

WB: 117", 114 (wagon)
OL: 206.2, 209 (wagon)
OW: 76.5 (4-dr exc. wagon); 76.8 (2-dr), 75.4 (wagon)
FT: 60.5
RT: 60
HT: 53.1 (4-dr sedan, hardtop), 52.3 (2-dr hardtop), 51.1 (SportsRoof); 52.7 (convertible), 55.7 (wagon)

MODELS

Series	Body Style	Model No.	Price
Torino	2-dr hardtop	65A	$2706
Torino 500	2-dr hardtop	65C	2887
Torino 500	2-dr hardtop (SportsRoof)	63C	2943
Torino Brougham	2-dr hardtop (V-8)	65E	3175
Torino GT	2-dr hardtop (SportsRoof, V-8)	63F	3150
Torino GT	2-dr convertible (V-8)	76F	3408
Torino Cobra	2-dr hardtop (SportsRoof, V-8)	63H	3295

(Prices as of 4/71)

ENGINES

Code	Type	CID	BxS	Hp@Rpm	Tq@Rpm	CR	Carb
L	I-6	250	3.68x3.91	145@4000	232@1600	9.0	1-bbl
F	V-8	302	4x3	210@4600	296@2600	9.0	2-bbl
H	V-8	351	4x3.5	240@4600	350@2600	9.0	2-bbl
M	V-8	351	4x3.5	285@5400	370@3400	10.7	4-bbl
C	V-8	429	4.36x3.59	370@5400	450@3400	11.3	4-bbl
J	V-8	429*	4.36x3.59	370@5400	450@3400	11.3	4-bbl

*Ram-Air

TRANSMISSIONS

3-Speed, All-synchro (Std. six, 302)
4-Speed (Opt. 351/285, 429)
Cruise-O-Matic (Opt. all, only trans. for 351/240)

SUSPENSION

Front: High coils and stabilizer bar
Rear: Leafs

PERFORMANCE MODELS

Torino GT—Models included 302/210; GT hood with integral non-functional scoop; rectangular GT badge in grille; dark argent rocker panel moldings with GT letters at base of front fender; full-width taillight effect with honeycomb applique and reflective center insert between taillights; black deck lid appliques (SportsRoof); color-keyed racing mirrors; hubcaps and wheel trim rings.

Torino Cobra—Model included 351/285; wide-ratio 4-speed; Hurst shifter; 7-inch-wide wheels; hubcaps; F70x14 raised-letter tires; flat-black hood; flat-black grille with chrome bar; heavy-duty load suspension; flat-black rear deck lid, panel, taillight bezels, fender extensions; wheel lip moldings; Cobra decals at rear of rear fenders, in center of panel between taillights.

OPTIONS

302/210, $95; 351/240, $140; 351/285, $188; 429/370, $374; 429/370 Ram-Air, $531; 4-Speed with Hurst shifter, $205; Cruise-O-Matic, $217 (six, 302, 351), $238 (429); Power front disc brakes, $70; Traction-Lok differential, $48; Heavy-duty load suspension, $23; Shaker hood scoop, $65 (351/285); Tachometer, 8000 rpm, $49; Styled steel wheels, argent, $58; Magnum 500 chrome wheels, $155; Bucket-type front seats, $150 (Torino 500 2-dr, GT, Cobra); Console, $60 (same models as buckets); Laser stripe, $39 (SportsRoof, GT convertible); Sports Slats, $65 (SportsRoof). (Prices as of 4/71)

1972 TORINO

DIMENSIONS

WB: 114" (2-dr), 118 (4-dr)
OL: 203.2 (Torino 2-dr hardtop), 207.7 (Torino 4-dr sedan), 211.5 (Torino wagon), 207.3 (Gran Torino 2-dr), 211.3 (Gran Torino 4-dr sedan), 215.1 (Gran Torino wagon)
OW: 79.3 (hardtop, sedan), 79 (wagon)
FT: 62.8, 63.4 (wagon)
RT: 62.9, 63.5 (wagon)
HT: 52.6 (sedan), 51.9 (hardtop), 51.8 (SportsRoof), 55 (wagon)

MODELS

Series	Body Style	Model No.	Price
Torino	2-dr hardtop*	65B	$2673
Gran Torino	2-dr hardtop*	65D	2878
Gran Torino Sport	2-dr hardtop (V-8)*	65R	3004
Gran Torino Sport	2-dr hardtop (SportsRoof, V-8)*	63R	3094

*Rallye Equipment Group optional (Prices as of 1/72)

ENGINES

Code	Type	CID	BxS	Hp@Rpm*	Tq@Rpm*	CR	Carb
L	I-6	250	3.68x3.91	95@3600	181@1600	8.0	1-bbl
F	V-8	302	4x3	140@4000	230@2200	8.5	2-bbl
H	V-8	351	4x3.5	161@4000	276@2000	8.6	2-bbl
Q	V-8	351	4x3.5	248@5400	299@3800	8.6	4-bbl
S	V-8	400	4x4	168@4200	297@2200	8.4	2-bbl
N	V-8	429	4.36x3.59	205@4400	322@2600	8.5	4-bbl

*Net ratings (Actual output of each engine varies per model)

TRANSMISSIONS

3-Speed, All-synchro (Std. six, 302)
4-Speed (Opt. 351/248)
Cruise-O-Matic (Opt. all, only trans. for 351/161, 400, 429)

SUSPENSION

Front: Conventional coils and stabilizer bar
Rear: Coils, four links (stabilizer bar with Rallye option)

PERFORMANCE MODELS

Gran Torino Sport—Models included 302/140; high-back bench seat; special interior door panels; nonfunctional hood scoop; color-keyed racing mirrors; hubcaps and wheel trim rings; wide-oval whitewall tires; color-keyed rear bumper pad; Gran Torino Sport nameplates on front fenders, behind wheels.

Rallye Equipment Group—Option included 351/248; 4-Speed; Hurst shifter; Instrumentation Group (tachometer, oil, amp, temperature and fuel gauges, trip odometer); competition suspension with rear stabilizer bar; 6-inch wheels; G70x14 belted, raised-letter tires.

OPTIONS

302/140, $95; 351/161, $140; 351/248, $225; 400/168, $194; 429/205, $327; 4-Speed with Hurst shifter, $205; Cruise-O-Matic, $217 (exc. 429), $238 (429); Power front disc brakes, $70; Bucket-type front seats, $150 (Gran Torino 2-dr hardtop, Gran Torino Sport); Console, $60 (same models as buckets); Instrumentation Group, $96; Traction-Lok differential, $48; Ram-Air hood scoop, $65 (Gran Torino hardtop or Gran Torino Sport models with 351/248 or 429/205); Rally Equipment Group, $453; Competition suspension, $31; Heavy-duty load suspension, $23; Magnum 500 chrome wheels, $155 (Torino, Gran Torino), $120 (Gran Torino Sport); Laser Stripe, $39 (2-dr exc. Torino). (Prices as of 1/72)

1973 TORINO

DIMENSIONS

WB: 114" (2-dr), 118 (4-dr)
OL: 208 (2-dr), 212 (4-dr sedan), 215.6 (wagon)
OW: 79.3 (hardtop, sedan), 79 (wagon)
FT: 62.8, 63.4 (wagon)
RT: 62.9, 63.5 (wagon)
HT: 52.9 (sedan), 52.2 (hardtop), 52.1 (SportsRoof), 54.9 (wagon)

MODELS

Series	Body Style	Model No.	Price
Torino	2-dr hardtop*	65B	$2673
Gran Torino	2-dr hardtop*	65D	2878
Gran Torino Sport	2-dr hardtop (V-8)*	65R	3094
Gran Torino Sport	2-dr hardtop (SportsRoof, V-8)*	63R	3094

*Rallye Equipment Group optional (Prices as of 12/72)

ENGINES

Code	Type	CID	BxS	Hp@Rpm*	Tq@Rpm*	CR	Carb
L	I-6	250	3.68x3.91	92@3200	197@1600	8.0	1-bbl
F	V-8	302	4x3	137@4200	230@2200	8.0	2-bbl
H	V-8	351	4x3.5	159@4000	260@2400	8.0	2-bbl
Q	V-8	351	4x3.5	246@5400	312@3600	8.0	4-bbl
S	V-8	400	4x4	168@3800	310@2000	8.0	2-bbl
N	V-8	429	4.36x3.59	201@4400	322@2600	8.0	4-bbl

*Net ratings (Actual output of each engine varies per model)

TRANSMISSIONS

3-Speed, All-synchro (Std. six, 302)
4-Speed (Opt. 351/246)
Cruise-O-Matic (Opt. all, only trans. for 351/159, 400, 429)

SUSPENSION

Front: Conventional coils and stabilizer bar
Rear: Coils, four links (stabilizer bar with Rallye option)

PERFORMANCE MODELS

Gran Torino Sport—Models included 302/137; special interior door panels; color-keyed racing mirrors; hubcaps and wheel trim rings; F70x14 raised-letter, wide-oval tires; wheel lip moldings; Gran Torino Sport nameplates on front fenders, behind wheels.

Rallye Equipment Group—Option included 351/246; 4-speed with Hurst shifter; Instrumentation Group; competition suspension with rear stabilizer bar; 6-inch wheels; G70x14 belted, raised-letter tires.

OPTIONS

302/137, $89-$91; 351/159, $133.96; 351/246, $216.87; 400/168, $186.59; 429/201, $316.76; 4-Speed, $199.78; Cruise-O-Matic, $210.89 (exc. 429), $231.26 (429); Power front disc brakes, $68.28; Bucket-type front seats, $146.14 (Gran Torino 2-dr hardtop, Gran Torino Sport); Console, $58.43 (same models as buckets); Instrumentation Group, $118.13; Traction-Lok differential, $46.83; Rallye Equipment Group, $385.37 (Gran Torino Sport), $442.37 (Torino, Gran Torino); Competition suspension, $30.17; Heavy-duty load suspension, $22.47; Magnum 500 chromed wheels, $151.07 (Torino, Gran Torino), $116.99 (Gran Torino Sport); Laser Stripe, $37.99 (2-dr except Torino). (Prices as of 12/72)

1965 MUSTANG

DIMENSIONS

WB: 108"
OL: 181.6
OW: 68.2
FT: 55.4 (six), 56 (V-8)
RT: 56
HT: 51.1 (hardtop), 51 (convertible), 51.2 (2+2)

MODELS

Series	Body Style	Model No.	Price
Mustang	2-dr hardtop	65A	$2372
Mustang	2-dr hardtop (2+2)*	63A	2589
Mustang	2-dr convertible	76A	2614

*Model introduced 9/64, others 4/64
GT Equipment Group optional all starting 2/65
(Prices as of 1/65)

ENGINES

Code	Type	CID	BxS	Hp@Rpm	Tq@Rpm	CR	Carb
U	I-6	170*	3.5x2.94	101@4400	156@2000	8.7	1-bbl
T	I-6	200#	3.68x3.13	120@4400	190@2400	9.2	1-bbl
F	V-8	260*	3.8x2.87	164@4400	258@2200	8.7	2-bbl
C	V-8	289#	4x2.87	200@4400	282@2400	9.3	2-bbl
D	V-8	289*	4x2.87	210@4400	300@2800	8.7	4-bbl
A	V-8	289#	4x2.87	225@4800	305@2800	10.0	4-bbl
K	V-8	289*	4x2.87	271@6000	312@3400	11.0	4-bbl
K	V-8	289#	4x2.87	271@6000	312@3400	10.9	4-bbl

*Available 4/64 through 9/64 #Available 9/64 on

TRANSMISSIONS

3-Speed (Std. six)
3-Speed, All-synchro (Std. 260, 289 exc. 271-hp)
4-Speed (Opt. all, only trans. for 289/271)
Cruise-O-Matic (Opt. exc. 289/271)

SUSPENSION

Front: High coils and stabilizer bar
Rear: Leafs

PERFORMANCE MODELS

GT Equipment Group—Option required 289/225 or 289/271. Option includes dual exhaust with flared, chrome tail-pipe extensions; heavy-duty suspension; 22:1 steering; front disc brakes; 4-inch fog lamps in grille; 5-dial instrument cluster; GT stripes above rocker panels; GT badges on front fenders, behind wheel and above stripes.

EMBLEMS

260 V-8—V-symbol with numbers 260 above it, on front of front fender.

289 V-8 (exc. 271-hp)—V-symbol with numbers 289 above and black below. On front of front fender.

289/271 V-8—V-symbol with numbers 289 above it and black underneath. Small chrome band over numbers with "High Performance" in capital letters, small checkered pattern alongside. On front of front fender.

OPTIONS

289/200, $108; 289/225, $162; 289/271, $442.60; 4-Speed, $115.90 (six), $188 (V-8); Cruise-O-Matic, $179.80 (six), $189.60 (V-8); Front disc brakes, $58; Limited-slip differential, $42.50; Special Handling Package, $31.30; Console, $51.50; Rally Pac, $70.80; Styled steel wheels, $122.30.

1966 MUSTANG

DIMENSIONS

WB: 108"

OL: 181.6
OW: 68.2
FT: 55.4 (six), 56 (V-8)
RT: 56
HT: 51.1 (hardtop), 51.2 (2+2), 51 (convertible)

MODELS

Series	Body Style	Model No.	Price
Mustang	2-dr hardtop	65A	$2398.43
Mustang	2-dr hardtop (2+2)	63A	2587.89
Mustang	2-dr convertible	76A	2633.34

GT Equipment Group optional all models

ENGINES

Code	Type	CID	BxS	Hp@Rpm	Tq@Rpm	CR	Carb
T	I-6	200	3.68x3.13	120@4400	190@2400	9.2	1-bbl
C	V-8	289	4x2.87	200@4400	282@2400	9.3	2-bbl
A	V-8	289	4x2.87	225@4800	305@3200	9.8	4-bbl
K	V-8	289	4x2.87	271@6000	312@3400	10.5	4-bbl

TRANSMISSIONS

3-Speed (Std. six)
3-Speed, All-synchro (Std. 289/200, 289/225)
4-Speed (Opt. all)
Cruise-O-Matic (Opt. all)

SUSPENSION

Front: High coils and stabilizer bar
Rear: Leafs

PERFORMANCE MODELS

GT Equipment Group—Option required 289/225 or 289/271. Option included dual exhaust with flared, chrome tail-pipe extensions; Special Handling Package; 22:1 steering; front disc brakes; 4-inch fog lights in grille; GT stripes above rocker panels; GT badges on front fenders, behind wheels and above stripes; GT fuel filler cap.

EMBLEMS

289 V-8 (exc. 271-hp)—V-symbol with numbers 289 above and black below. On front of front fender.
289/271 V-8—V-symbol with numbers 289 above it and black underneath. Small chrome band over number with "High Performance" in capital letters, small checkered pattern alongside. On front of front fender.

OPTIONS

289/200, $104.84; 289/225, $157.31; 289/271, $430.39; 4-Speed, $112.63 (six), $183.99 (V-8); Cruise-O-Matic, $174.47 (six), $183.99 (289/200, 289/225), $214.63 (289/271); Front disc brakes, $56.36; Limited-slip differential, $41.30; Special Handling Package, $30.42; Console, $50.05; Rally Pac, $68.80; Styled steel wheels, $93.16; GT Equipment Group, $151.10; Wire wheel covers, $57.82; Full-width front seat, $24.24 (exc. 2+2).

1967 MUSTANG

DIMENSIONS

WB: 108"
OL: 183.6
OW: 70.9
FT: 58
RT: 58
HT: 51.6 (hardtop, convertible), 51.8 (2+2)

MODELS

Series	Body Style	Model No.	Price
Mustang	2-dr hardtop	65A	$2461.46
Mustang	2-dr hardtop (2+2)	63A	2592.17
Mustang	2-dr convertible	76A	2698.14

GT Equipment Group optional all models, GTA with automatic
(Prices as of 10/66)

ENGINES

Code	Type	CID	BxS	Hp@Rpm	Tq@Rpm	CR	Carb
T	I-6	200	3.68x3.13	120@4400	190@2400	9.2	1-bbl
C	V-8	289	4x2.87	200@4400	282@2400	9.3	2-bbl
A	V-8	289	4x2.87	225@4800	305@3200	9.8	4-bbl
K	V-8	289	4x2.87	271@6000	312@3400	10.5	4-bbl
S	V-8	390	4.05x3.78	320@4800	427@3200	10.5	4-bbl

TRANSMISSIONS

3-Speed, All-synchro (Std. exc. 289/271)
4-Speed (Opt. V-8)
Cruise-O-Matic (Opt. all)

SUSPENSION

Front: High coils and stabilizer bar
Rear: Leafs

PERFORMANCE MODELS

GT Equipment Group—Option required a V-8. Option included Special Handling Package; power front disc brakes; 4-inch fog lights in grille; GT stripes above rocker panel; GT or GTA (GTA with automatic) within stripe on front fender; GT fuel filler cap; chromed quad exhaust extensions if equipped with 289/271 or 390/320.

EMBLEMS

289 V-8—Mustang emblem on side of car, behind front wheel well had red, white and blue vertical rectangle behind it, at top are numbers 289. Not on GT-optioned cars.
390 V-8—Mustang emblem on side of car, behind front wheel well had red, white and blue vertical rectangle behind it, at top are numbers 390. Not on GT-optioned cars.

OPTIONS

289/200, $105.63; 289/225, $158.48; 289/271, $433.55 (with GT Group only); 390/320, $263.71; 4-Speed, $184.02 (289/200, 289/225); $233.18; 289/271, 390/320; Cruise-O-Matic, $188.18 (six), $197.89 (289/200, 289/225), $220.17 (289/271, 390/320); Heavy-duty 3-speed, $79.20 (required 390/320); Power front disc brakes, $64.77; Limited-slip differential, $41.60; Competition Handling package, $388.53 (with GT Group only); Special Handling Package, NA; Console, $50.41; Styled steel wheels, $115.11 (hardtop, conv.), $93.84 (2+2); GT Equipment Group, $205.05; Wire wheel covers, $79.51 (hardtop, conv.), $58.24 (2+2); Wide-oval tires, $62.35 (V-8). (Prices as of 10/66)

1968 MUSTANG

DIMENSIONS

WB: 108"
OL: 183.6
OW: 70.9
FT: 58.5
RT: 58.5
HT: 51.6 (hardtop, 2+2), 51.4 (convertible)

MODELS

Series	Body Style	Model No.	Price
Mustang	2-dr hardtop	65A	$2647.78
Mustang	2-dr hardtop (2+2)*	63A	2758.44
Mustang	2-dr convertible	76A	2861.47

*Cobra Jet package, must have GT package (mid-year)
GT Equipment Group optional all models
Sports Trim Group optional all models (discontinued 12/67)
(Prices as of 2/68)

ENGINES

Code	Type	CID	BxS	Hp@Rpm	Tq@Rpm	CR	Carb
T	I-6	200	3.68x3.13	115@3800	190@2200	8.8	1-bbl
C	V-8	289	4x2.87	195@4600	288@2600	8.7	2-bbl
J	V-8	302	4x3	230@4800	310@2800	10.0	4-bbl
X	V-8	390*	4.05x3.78	280@4400	403@2600	10.5	2-bbl
S	V-8	390	4.05x3.78	325@4800	427@3200	10.5	4-bbl
W	V-8	427#	4.23x3.78	390@5600	460@3200	10.9	4-bbl
R	V-8	428*	4.13x3.98	335@5400	440@3400	10.6	4-bbl

*Mid-year engine #Discontinued mid-year

TRANSMISSIONS

3-Speed, All-synchro (Std. six, 289, 302)
4-Speed (Opt. exc. six, 427)
Cruise-O-Matic (Opt. all, only trans. for 390/280, 427/390)

SUSPENSION

Front: High coils and stabilizer bar
Rear: Leafs

PERFORMANCE MODELS

GT Equipment Group—Option required 302/230, 390/280, 390/325, 427/390 or 428/335. Option included heavy-duty suspension; 4-inch fog lights in grille; C-stripe on sides of cars (old-style stripes above rocker panels also available); GT emblem on front fender, behind wheel well; GT fuel filler cap; styled steel wheels with GT hubcaps; F70x14 wide-oval white stripe tires; chromed quad exhaust extensions.
Sports Trim Group—Option included black stripes on hood (required louvered hood); wheel lip moldings; wood-grain instrument panel applique; argent styled steel wheels with plain hubcaps (V-8); E70x14 wide-oval white-band tires (V-8). (Option discontinued 12/67, but items available individually after that.)
Cobra Jet—Option available on 2+2 with GT Equipment Group. Option included 428/335; functional hood scoop; flat-black stripe down center of hood to base of windshield (included scoop); power front disc brakes; staggered rear shocks (4-speed-equipped cars); 8000-rpm tachometer (4-speeds); 9-inch differential ring gear.

EMBLEMS

302 V-8—Mustang horse emblem on side of car, behind front wheel well, had red, white and blue vertical rectangle behind it, at top are numbers 302. Not on GT-optioned cars.
390 V-8—Mustang horse emblem on side of car, behind front wheel well, had red, white and blue vertical rectangle behind it, at top are numbers 390. Not on GT-optioned cars.

OPTIONS

289/195, $105.63; 302/230, $171.77; 390/280, $206.73; 390/325, $263.71; 4-Speed, $184.02 (289, 302), $233.18 (390/325, 428/335); Cruise-O-Matic, $191.13 (six), $200.85 (289, 302), $223.03 (390/280), $233.17 (390/225, 427/390, 428/335); Power front disc brakes, $64.77; Limited-slip differential, $41.60; Console, $53.17; GT Equipment Group, $146.71; Tachometer and trip odometer, $54.45; F70x14 wide-oval tires, $78.53. (Prices as of 2/68)

1969 MUSTANG

DIMENSIONS

WB: 108"
OL: 187.4
OW: 71.3, 71.8 (Boss 302), 71.9 (Boss 429)
FT: 58.5, 59.3 (Boss 429), 59.5 (Boss 302)
RT: 58.5, 59.5 (Boss 302, Boss 429)
HT: 51.3 (hardtop), 51.5 (convertible), 50.6 (SportsRoof), 50.4 (Mach 1, Boss 302, Boss 429)

MODELS

Series	Body Style	Model No.	Price
Mustang	2-dr hardtop*	65A	$2635
Mustang	2-dr sedan (SportsRoof)*†#	63A	2635
Mustang	2-dr convertible*	76A	2849
Grande	2-dr hardtop	65E	2866
Mach 1	2-dr sedan (SportsRoof, V-8)	63C	3122

*GT Equipment Group optional ($146.71)
†Boss 302 package optional (mid-year)
#Boss 429 package optional (mid-year) ($3948, $4789 with mandatory options)
(Prices as of 2/69)

ENGINES

Code	Type	CID	BxS	Hp@Rpm	Tq@Rpm	CR	Carb
T	I-6	200	3.68x3.13	115@3800	190@2200	8.8	1-bbl
L	I-6	250	3.68x3.91	155@4000	240@1600	9.0	1-bbl
F	V-8	302	4x3	220@4600	300@2600	9.5	2-bbl
G	V-8	302*	4x3	290@5800	290@4300	10.5	4-bbl
H	V-8	351	4x3.5	250@4600	355@2600	9.5	2-bbl
M	V-8	351	4x3.5	290@4800	385@3200	10.7	4-bbl
S	V-8	390	4.05x3.78	320@4600	427@3200	10.5	4-bbl
Q	V-8	428	4.13x3.98	335@5200	440@3400	10.6	4-bbl
R	V-8	428#	4.13x3.98	335@5200	440@3400	10.6	4-bbl
Z	V-8	429*	4.36x3.59	375@5200	410@3400	10.5	4-bbl

*Mid-year engine #Ram-Air

TRANSMISSIONS

3-Speed, All-synchro (Std. six, 302/220, 351)
4-Speed (Opt. exc. six, only trans. for Boss 302 and Boss 429)
Cruise-O-Matic (Opt. exc. Boss 302 and Boss 429)

SUSPENSION

Front: High coils and stabilizer bar
Rear: Leafs (and stabilizer bar on Boss 429)

PERFORMANCE MODELS

GT Equipment Group—Option required 351, 390 or 428. Option included heavy-duty suspension; stripes parallel to rocker panel, but without emblems; GT fuel filler cap; nonfunctional hood scoop with rear turn signal indicators; hood pins; styled steel wheels with GT hubcaps; E70x14 white-stripe tires; chrome quad exhaust extensions on cars with dual exhaust.
Mach 1—Model included 351/250; heavy-duty suspension; flat-black panels on hood and cowl; hood pins; nonfunctional hood scoop with turn signal indicators; stripe on side of car starting with name reversed out, continuing to rear marker light, interrupted by rear wheel opening; rear deck stripe with name reversed out; pop-open fuel filler cap; styled steel wheels with plain hubcaps; E70x14 white-stripe tires; chrome rocker panel trim; Mach 1 nameplate with checkered pattern on right on panel over glove compartment; rally clock; simulated wood applique on instrument panel and console; high-back bucket-type seats; extra insulation; bright-trimmed pedals.
Boss 302—Option included 302/290; 4-speed; Competition suspension; power front disc brakes; 16:1 steering; Magnum 500 argent, 7-inch wheels; F60x15 raised-letter tires; flat-black panels on hood and cowl; front spoiler; flat-black paint around headlights; C-stripes on sides, starting behind front wheel opening with Boss 302 lettering at front; flat-black paint on rear deck; flat-black paint on panel under rear deck; rear quarter panel scoops filled in; flared front fenders.
Boss 429—Option included 429/375; 4-speed; Competition suspension with revised front suspension and rear stabilizer bar; power front disc brakes; power steering; engine oil cooler; Traction-Lok differential with 3.91:1 gears; trunk-mounted battery; functional hood scoop; front spoiler; Boss 429 decals on front fender, behind wheel opening; Magnum 500 chrome 7-inch wheels; F60x15 raised-letter tires; Interior Decor Group with high-backed bucket-type seats and console; color-keyed racing mirrors; flared front fenders (some of these items were mandatory options).

EMBLEMS

351 V-8—If car had no scoop a small parallelogram, leaning toward the right with the numbers 351 against a black background was mounted below the Mustang nameplate on the front fenders, behind the wheel opening. If it had a scoop, a rectangle with a ribbed background and the numbers 351 was on either side of the scoop.
390 V-8—If car had no scoop a small parallelogram, leaning toward the right with the numbers 351 against a black background was mounted below the Mustang nameplate on the front fenders, behind the wheel opening. If it had a scoop, a rectangle with a ribbed background and the numbers 390 were on either side of the scoop.
428 V-8—Cars with scoops got Cobra Jet in a script nameplate on either side of scoop.

OPTIONS

250/155, $25.91; 302/220, $105.00; 351/250, $163.34; 351/290, $189.25; 390/320, $163.08; 428/335, $392.53; 428/335 Ram-Air, $525.96; 4-speed, $204.64 (302/220, 351); $253.92 (390, 428); Cruise-O-Matic, $191.13 (six); $200.85 (302/220, 351); $222.08 (390, 428); Power front disc brakes, $64.77; Limited-slip differential, $41.60 (250, 302/220); Traction-Lok differential, $63.51 (302/290, 351, 390, 428); Functional hood scoop, $45.39 (351), $84.25 (390); GT Equipment Group, $146.71; Tachometer, $54.45; Color-keyed racing mirrors, $19.48; Special Handling Package, $30.64 (302/220, 351, 390, NA Grande); Competition suspension, $30.64; Console, $53.82; Chrome styled steel wheels, $116.59; Argent styled steel wheels, $38.86; Wire wheel covers, $79.51. (Prices as of 2/69)

1970 MUSTANG

DIMENSIONS

WB: 108"
OL: 187.4
OW: 71.7, 71.8 (Boss 429)
FT: 58.5, 59.5 (Boss 302, Boss 429)
RT: 58.5, 59.5 (Boss 302, Boss 429)
HT: 51.5 (hardtop, convertible), 50.6 (SportsRoof), 50.4 (Mach 1, Boss 429); 50.2 (Boss 302)

MODELS

Series	Body Style	Model No.	Price
Mustang	2-dr hardtop	65A	$2721
Mustang	2-dr sedan (SportsRoof)*†	63A	2771
Mustang	2-dr convertible	76A	3025
Grande	2-dr hardtop	65E	2926
Mach 1	2-dr sedan (SportsRoof, V-8)	63C	3271

*Boss 302 package optional brings price to $3720.
†Boss 429 package optional, brings price to $4932 with mandatory options. Discontinued mid-year.
(Prices as of 12/19/69)

ENGINES

Code	Type	CID	BxS	Hp@Rpm	Tq@Rpm	CR	Carb
T	I-6	200	3.68x3.13	120@4000	190@2200	8.7	1-bbl
L	I-6	250	3.68x3.91	155@4000	240@1600	9.0	1-bbl
F	V-8	302	4x3	220@4600	300@2600	9.5	2-bbl
G	V-8	302	4x3	290@5800	290@4300	10.6	4-bbl
H	V-8	351	4x3.5	250@4600	335@2600	9.5	2-bbl
M	V-8	351	4x3.5	300@5400	380@3400	11.0	4-bbl
Q	V-8	428	4.13x3.98	335@5200	440@3400	10.6	4-bbl
R	V-8	428*	4.13x3.98	335@5200	440@3400	10.6	4-bbl
Z	V-8	429†	4.36x3.59	375@5600	450@4000	10.5	4-bbl

*Ram-Air †Discontinued mid-year

TRANSMISSIONS

3-Speed, All-synchro (Std. six, 302/220, 351)
4-Speed (Opt. exc. six, only trans. for Boss 302, Boss 429)
Cruise-O-Matic (Opt. exc. Boss 302, Boss 429)

SUSPENSION

Front: High coils and stabilizer bar
Rear: Leafs (and stabilizer bar on Boss 302, Boss 429, Mach 1)

PERFORMANCE MODELS

Mach 1—Model included 351/250; Competition suspension with rear stabilizer bar; flat-black or flat-white panel on hood with stripes and numbers for displacement size; round external hood latches; nonfunctional hood scoop with turn signal indicators in rear; aluminum rocker panel moldings with Mach 1 lettering cast into panel for lower front fender; rear deck and quarter-panel end stripe, breaking for Mach 1 chrome letters; honeycomb pattern applique for rear panel between taillights; pop-open fuel filler cap; simulated mag wheel covers; E70x14 raised-letter tires; grille with driving lights; Mach 1 nameplate on panel over glove compartment; rally clock; high-back bucket-type seats; console; simulated wood-grain steering wheel, instrument panel and console appliques; bright-trimmed pedals.
Boss 302—Option included 302/290; 40speed with Hurst shifter; Competition suspension with rear stabilizer bar; power front disc brakes; 16:1 steering; 7-inch wheels with hubcaps and trim rings; F60x15 raised-letter tires; flat-black panel on hood with stripes on either side that turn at back at right angles and go onto fender; side flat-black stripes that start on fender below stripe from hood with Boss 302 lettering, angle down, then go to back of car; flat-black paint on rear deck; black panel between taillights and taillight bezels; flared front fenders; front spoiler.
Boss 429—Option included 429/375; 4-speed; Competition suspension with revised front suspension and rear stabilizer bar; power front disc brakes; power steering with 16:1 ratio; engine oil cooler; Traction-Lok differential with 3.91:1 gears; trunk-mounted battery; functional hood scoop, black only; front spoiler; Boss 429 decals on front fender, behind wheel openings; Magnum 500 chrome 7-inch wheels; F60x15 raised-letter tires; Interior Decor Group with high-back bucket-type seats, console; clock; color-keyed racing mirrors; flared front fenders (some of these items were mandatory options for the package).

EMBLEMS

351 V-8—If car had no scoop, a small parallelogram, leaning toward the right with the numbers 351 against a black background was mounted below the Mustang nameplate on the front fenders, behind the wheel opening. Mach 1 models had the numbers 351 in tape in a break in the stripe on the hood on either side of the scoop.
428 V-8—Cars with scoops got Cobra Jet in a script nameplate on either side of the scoop. Mach 1 models got the numbers 428 in tape in a break in the stripe next to the hood on either side of the scoop.

OPTIONS

250/155, $39; 302/220, $101; 351/250, $146; 351/300, $194; 428/335, $457; 428/335 Ram-Air, $522; 4-speed, $205; Cruise-O-Matic, $201 (six, 302/220, 351), $222 (428); Power front disc brakes, $65; Traction-Lok differential, $43; Drag Pack with 3.91:1 Traction-Lok differential, $155 (428 only); Drag Pack with 4.30:1 Detroit Locker differential, $207 (428 only); Shaker hood scoop, $65 (Boss 302, 351/250, 351/300); Rear deck spoiler, $20 (SportsRoof, Mach 1, Boss 302, Boss 429); Tachometer and trip odometer, $54; Console, $54; Color-keyed racing mirrors, $26; Sport Slats, $65; Wire wheel covers, $79; Magnum 500 chrome wheels, $129 (Boss 302); Styled steel wheels, argent, $58 (no charge on Mach 1, Boss 302); Quick-ratio (16:1) steering, $16; Heavy-duty cooling system, $13. (Prices as of 12/19/69)

1971 MUSTANG

DIMENSIONS

WB: 109"
OL: 189.5
OW: 74.1
FT: 61.5
RT: 61
HT: 50.8 (hardtop), 50.5 (convertible), 50.1 (SportsRoof)

MODELS

Series	Body Style	Model No.	Price
Mustang	2-dr hardtop#	65D	$2911

Mustang	2-dr hardtop (SportsRoof)*	63D	2973
Mustang	2-dr convertible	76D	3227
Grande	2-dr hardtop	65F	3117
Mach 1	2-dr hardtop (SportsRoof, V-8)	63R	3268

*Boss 351 option package brings price to $4124.
#Special Value Package (mid-year) brings price to $3008.40.
(Prices as of 4/71)

ENGINES

Code	Type	CID	BxS	Hp@Rpm	Tq@Rpm	CR	Carb
L	I-6	250	3.68x3.91	145@4000	232@1600	9.0	1-bbl
F	V-8	302	4x3	210@4600	296@2600	9.0	2-bbl
H	V-8	351	4x3.5	240@4600	350@2600	9.0	2-bbl
Q	V-8	351*	4x3.5	280@5800	345@3800	9.0	4-bbl
M	V-8	351	4x3.5	285@5400	370@3400	10.7	4-bbl
R	V-8	351	4x3.5	330@5400	370@4000	11.7	4-bbl
C	V-8	429	4.36x3.59	370@5400	450@3400	11.3	4-bbl
J	V-8	429†	4.36x3.59	370@5400	450@3400	11.3	4-bbl
C**	V-8	429	4.36x3.59	375@5600	450@4000	11.3	4-bbl
J**	V-8	429†	4.36x3.59	375@5600	450@4000	11.3	4-bbl

*Mid-year engine †Ram-Air **Axle code has V (3.91) or Y (4.11)

TRANSMISSIONS

3-Speed, All-synchro (Std. six, 302, 351/240)
4-Speed (Opt. 351 exc. 240-hp, 429, only trans. for Boss 351)
Cruise-O-Matic (Opt. exc. Boss 351)

SUSPENSION

Front: High coils and stabilizer bar
Rear: Leafs

PERFORMANCE MODELS

Mach 1—Model included 302/210; Competition suspension; standard flat hood, but optional at no charge for 302-powered cars was a hood with twin, nonfunctional scoops, with 351 or 429 engines, scooped hood was standard; flat-black or argent paint at bottom of front fenders, doors, on rocker panels and bottom of rear fenders with chrome strip on top; Mach 1 decals on front fenders, behind wheel, over painted area; honeycomb pattern grille with driving lights; color-keyed front bumper; honeycomb-pattern black applique on panel between taillights; stripe on rear deck and quarter panel ends; pop-open fuel filler cap; hubcaps and wheel trim rings; E70x14 white-stripe tires; Mach 1 plate in center of instrument panel; color-keyed racing mirrors.
Boss 351—Option included 351/330; 4-speed with Hurst shifter; Competition suspension; power front disc brakes; 7-inch wheels with hubcaps and wheel trim rings; F60x15 raised-letter tires; Mach 1 grille and driving lights; color-keyed or chrome front bumper; flat-black or argent on most of hood, which had functional twin scoops, round external latches; Ram-Air decals on outside of each scoop; lower body paint in flat-black or argent; stripes from front-marker light to below body line, then to rear of car, interrupted by wheel openings; Boss 351 decals on each front fender behind wheel opening and between stripe and lower paint; black or argent lower back panel; Boss 351 decal in center of rear deck lid; front spoiler; Instrumentation Group (included four gauges, speedometer, trip odometer, tachometer); heavy-duty cooling package; 3.91:1 Traction-Lok differential; color-keyed racing mirrors.

EMBLEMS

351 V-8—When functional Ram-Air hood used, 351 numbers in decals were next to Ram-Air lettering on scoop. Not on all.
429 V-8—When functional Ram-Air hood used, 429 numbers in decals were next to Ram-Air lettering on scoop. Not on all.

OPTIONS

302/210, $95; 351/240, $140; 351/285, $188; 429/370, $467; 429/370 Ram-Air, $531; 4-speed with Hurst shifter, $216; Cruise-O-Matic, $217 (six, 302, 351 exc. Boss), $238 (429); Power front disc brakes, $70; Traction-Lok differential, $48; Drag Pack with 3.91:1 Traction-Lok differential, $155 (429 only); Drag Pack with 4.11 Detroit Locker differential, $207 (429 only); Ram-Air hood with functional scoops, $65 (351/240, 351/285); Rear deck spoiler, $32 (SportsRoof, Mach 1, Boss 351); Instrumentation Group, $79; Console, $76; Color-keyed racing mirrors, $26; Boss 351-type tape stripes for side of car, $26; Magnum 500 chrome wheels, $155; Heavy-duty cooling system; Competition suspension, $31; Power windows, $127. (Prices as of 4/71)

1972 MUSTANG

DIMENSIONS

WB: 109"
OL: 189.5
OW: 74.1
FT: 61.5
RT: 61
HT: 50.8 (hardtop), 50.5 (convertible), 50.1 (SportsRoof)

MODELS

Series	Body Style	Model No.	Price
Mustang	2-dr hardtop*	65D	$2729
Mustang	2-dr sedan (SportsRoof)*†	63D	2786
Mustang	2-dr convertible	76D	3015
Grande	2-dr hardtop	65F	2915
Mach 1	2-dr sedan (SportsRoof, V-8)†	63R	3053

*Sprint option package
†Power windows option makes a 2-dr hardtop
(Prices as of 1/72)

ENGINES

Code	Type	CID	BxS	Hp@Rpm	Tq@Rpm	CR	Carb
L	I-6	250	3.68x3.91	99@3600	164@1600	8.0	1-bbl
F	V-8	302	4x3	141@4000	242@2000	8.5	2-bbl
H	V-8	351	4x3.5	177@4000	284@2000	8.6	2-bbl
Q	V-8	351	4x3.5	266@5400	301@3600	8.6	4-bbl
R	V-8	351#	4x3.5	275@6000	286@3800	9.2	4-bbl

*Net ratings #Mid-year engine

TRANSMISSIONS

3-Speed, All-synchro (Std. six, 302, 351/177)
4-Speed (Opt. 351/266, 351/275)
Cruise-O-Matic (Opt.)

SUSPENSION

Front: High coils and stabilizer bar
Rear: Leafs

PERFORMANCE MODELS

Mach 1—Model included 302/141; Competition suspension; standard flat hood, but optional at no charge for 302-powered cars was a hood with twin, nonfunctional scoops, with 351, scooped hood was standard; flat-black or argent paint at bottom of front fenders, doors, rocker panels, rear fenders with chrome strip on top; Mach 1 decals on front fenders, behind wheels, over painted area; honeycomb-pattern grille with driving lights; color-keyed front bumper; honeycomb-pattern black applique on panel between taillights; stripe on rear deck and quarter panel ends; hubcaps and wheel trim rings; E70x14 white-stripe tires; Mach 1 plate in center of instrument panel; color-keyed racing mirrors.

EMBLEMS

351 V-8—When functional Ram-Air hood used, 351 numbers in decals next to Ram-Air lettering on scoop.

OPTIONS

302/141, $90-$91; 351/177, $132; 351/266, $209; 351/275, $985 (includes 4-speed, etc.); 4-speed with Hurst shifter, $199; Cruise-O-Matic, $210; Power front disc brakes, $64; Traction-Lok differential, $44; Ram-Induction hood, $60 (351); Rear deck spoiler, $30 (SportsRoof, Mach 1); Instrumentation Group, $73; Console, $70; Color-keyed racing mirrors, $24; Tape stripes like 1971 Boss 351, $24; Magnum 500 chrome wheels, $143; Competition suspension, $29; Power windows, $117. (Prices as of 10/27/71)

1973 MUSTANG

DIMENSIONS

WB: 109"
OL: 193.8
OW: 74.1
FT: 61.5
RT: 61
HT: 50.7 (hardtop), 50.4 (convertible), 50 (SportsRoof)

MODELS

Series	Body Style	Model No.	Price
Mustang	2-dr hardtop	65D	$2760
Mustang	2-dr sedan (SportsRoof)*	63D	2820
Mustang	2-dr convertible	76D	3102
Grande	2-dr hardtop	65F	2946
Mach 1	2-dr sedan (SportsRoof, V-8)*	63R	3088

*Power window option makes a 2-dr hardtop
(Prices as of 1/73)

ENGINES

Code	Type	CID	BxS	Hp@Rpm*	Tq@Rpm*	CR	Carb
L	I-6	250	3.68x3.91	95@3200	199@1600	8.0	1-bbl
F	V-8	302	4x3	141@4000	229@2600	8.0	2-bbl
H	V-8	351	4x3.5	173@4400	258@2400	8.0	2-bbl
Q	V-8	351	4x3.5	259@5600	292@3400	7.9	4-bbl

*Net ratings

TRANSMISSIONS

3-Speed, All-synchro (Std. exc. 351/259)
4-Speed (Opt. 351/259)
Cruise-O-Matic (Opt. all)

SUSPENSION

Front: High coils and stabilizer bar
Rear: Leafs

PERFORMANCE MODELS

Mach 1—Model included 302/141; Competition suspension; standard flat hood, but optional at no extra charge for 302-powered cars was a hood with twin, nonfunctional scoops, with 351 scooped hood was standard; stripes on sides starting ahead of front side-marker light and extending back to rear quarter panel in front of rear wheel with Mach 1 reversed out on rear quarter, stripe broken for front wheel opening; honeycomb-pattern grille with flat-black headlight bezels; honeycomb-pattern black applique on panel between taillights; stripe on rear deck and quarter panel ends, broken on right side for Mach 1 decal; hubcaps with wheel trim rings; E70x14 white-stripe tires; Mach 1 plate in center of instrument panel; color-keyed racing mirrors.

EMBLEMS

351 V-8—When 351/173 came with optional functional Ram-Air hood, decals on outer sides read 351 Ram-Air. (351/173 was only engine available with Ram-Air hood.)

OPTIONS

302/141, $87; 351/173, $127.79; 351/259, $214; 4-speed with Hurst shifter, $192.99; Cruise-O-Matic, $203.73; Power front disc brakes, $62.05; Traction-Lok differential, $42.64; Ram-Induction hood, $58.24 (351/173 only); Rear deck spoiler, $29.12 (SportsRoof, Mach 1); Instrumentation Group, $70.83; Console, $67.95; Color-keyed racing mirrors, $23.23; Tape stripes like 1971 Boss 351, $23.23; Forged aluminum wheels, $142; Competition suspension, $28.19; Power windows, $113.48. (Prices as of 1/73)

INDEX

AA Gas Willys body, 150
AC models, 23, 86, 136, 143
Acme car, 6
Adamowicz, Tony, 144
Alaska-Yukon-Pacific Exposition, 6
Alfa Romeo models, 144
Allen, Johnny, 39
Allison, Bobby, 106, 113, 121, 130
Allison, Donnie, 113, 122, 123, 130, 140
Amarillo Dragway, 31
American 500 race, 75, 100, 122, 130
American Hot Rod Association (AHRA), 42, 94
American Manufacturers Association (AMA), 9, 10, 11, 12, 13, 14, 35-36, 50-51, 52, 57, 112, 124, 172
American Motors models, 153
American Racing Associates, 166
American Sunroof Corporation, 174
Amon, Chris, 82
Andretti, Mario, 82, 104, 105-106, 113
Arcaro, Eddie, 37
Arrow models, 6
Atlanta 250 race, 45
Atlanta 500 race, 48, 49-50, 64, 105-106, 114
Austin models, 27
Auto Competition Committee of the U.S. (ACCUS), 75
Autolite, 44, 154, 163, 169
Auto Racing Club of America (ARCA), 65, 70, 110, 114
Auto Sports, 54

Baker, Buddy, 122
Barker, John, 60
Beauchamp, Johnny, 14
Bendix brakes, 25
Bettenhausen, Tony, 45
Black Pearl show car, 73
Bohl, Cathy, 161
Bohl, Roger, 161
Bondy Long Ford, 61, 68, 103, 113
Bonner, Phil, 28, 29, 142-143
Bonneville Salt Flats, 9, 21, 53, 160
Borg-Warner transmission, 22, 24, 36, 41, 45, 46, 52, 68, 69, 88, 91, 135, 145
Bowsher, Jack, 65, 70, 74, 104, 106, 113-114, 116, 118, 123, 130, 131
Brannan, Dick, 141, 142
Brickhouse, Richard, 123
Bristol Dragway, 70, 73
Brock, Ray, 37, 138, 160
Brooker, Brad, 144
Bryar Motorsports Park, 165, 166
Bucknum, Ronnie, 149, 161
Buick models, 8, 24, 98, 107, 128
Burke, Bill, 21
Buyer's Digest, 81

Car and Driver, 25, 27, 157, 160, 171
Car Life, 48, 69, 73, 81, 86, 98, 138, 145, 159
Carlsson, Erik, 25
Carter carburetor, 27
Centro per L'Alta Moda Italiana, 42
Chapman, Colin, 135
Cheesbourg, Bill, 51
Chevrolet models, 12, 20, 21, 22, 24, 43, 45, 98, 107, 114, 117, 128, 135, 136, 137, 143, 145, 146, 147-148, 150, 151, 156, 160, 163, 165, 169, 172, 175
Chrisman, Jack, 106
Chrysler models, 7, 8, 9, 27, 35, 144, 146
Clark, Jim, 135
Coleman, Milo, 88
Continental Divide Raceway, 20, 149
Coons, Bill, 17
Cowan, Andrew, 142
Cowley, John, 36
Crawford, Ed, 21
Crow's Landing racetrack, 149
Cunningham, Briggs, 21

Dahlquist, Eric, 152-153
Darlington International Speedway, 8, 40, 45, 49, 63, 114, 122
Daytona 500 race, 14, 34, 38, 44, 63, 70, 105, 112, 113, 122, 123, 130, 150
Daytona Grand National, 10
Daytona International Speedway, 13-14, 21, 37, 38, 40, 42, 49, 53, 56, 57, 58, 62, 63, 68, 75, 105, 115, 120, 149
Dearborn Steel Tubing Company, 23, 28, 52, 59, 90
DeLorean, George, 143
DeLorenzo, Tony, 166
Detroit Automotive Products, 55
Detroit Locker differential, 90
Devin car, 59
Dibos, Eduardo, 14
Dieringer, Darel, 58, 62, 64, 66, 75, 100, 104
Dixie 400 race, 58
Dixie 500 race, 122
Dodge models, 24, 29, 47, 52, 70, 74, 86, 92, 105, 107, 117-118, 121, 122, 123, 129, 135, 136, 138, 144, 149, 163
Donnybrooke racetrack, 161, 165
Donohue, Mark, 149, 153, 160, 165-166
Droke, Darrell, 94
Duesenberg Special, 53
Dunlop SP tires, 25

East African Safari Rally, 33
Eckstrand, Al, 52, 59
Edelbrock Equipment Company, 20
Elford, Vic, 166
Elliott, Bill, 18
Evans, Dave, 36, 37

Fagersta Bruk tires, 25
Federation International de l'Automobile (FIA), 82
Firecracker 250, 14
Firecracker 400 race, 58, 64, 68, 122
Firestone tires, 140
Fischer, Craig, 153
Flock, Tim, 14
Follmer, George, 160-161, 165, 166
Ford, Benson, 25
Ford, Bob, 145
Ford, Henry II, 6, 50, 75, 84, 129
Ford, William Clay, 22
Ford Dealer, 121
Ford Motor Company
  Atlanta plant, 121
  Dallas, Texas, assembly plant, 41
  Design Center, 78, 108
  Kar Kraft, 118, 158, 164
  Romeo, Michigan, Proving Grounds, 36, 53
  San Jose assembly plant, 143
  Special Vehicles Department, 88
  Wixom, Michigan, plant, 11
Fox, Ray, 57-58
Foyt, Anthony Joseph "A. J.," 51, 57, 63, 64, 68, 70, 82, 104, 106, 113-114, 118, 120-123, 130, 135
France, Bill, 45
Frank, Larry, 49
Gapp, Wayne, 135
Gerken, Whitey, 106
Glidden, Bob, 135
Glotzbach, Charlie, 122, 130
Goldsmith, Paul, 40, 45, 51, 58, 63
Gonzales, Tubby, 44-45
Great Lakes Dragaway, 31
Green Valley Raceway, 28, 144, 149
Gregg, Peter, 166
Gregson, Dean, 147
Grosz, Jerry, 92
Grove, Tom, 59, 143
Guggenheim, M. Robert, 6
Gurney, Dan, 56-57, 62, 70, 74, 82, 104, 113, 120, 122, 130, 135, 149, 153, 165

Haggbom, Gunnar, 25
Hall, Anne, 25, 28
Hall, Jim, 23, 165-166
Hamilton, Pete, 130
Hane, Walter, 146
Hansgen, Walt, 20
Hanyon, Bill, 59
Harper, H. B., 6
Harper, Peter, 139
Harvey, Jerry, 145, 153
Harvey, Scott, 144
Hedrick, Ed, 136
Hertz Rent-A-Car, 145-146
Hilborn fuel injection, 21
Hoefer, Bill, 70
Hoerr, Rudy, 113, 130
Holley carburetor, 20, 36, 41, 43, 53, 55, 63, 115 127, 149, 154, 158
Holman, John, 9, 21
Holman & Moody, 13, 14, 20, 21, 23, 25, 29, 37, 39, 40, 44, 45, 48-49, 50, 53, 56, 58, 63, 64, 68-69, 70, 74, 75, 82, 100, 104, 105-106, 113, 120, 122, 123, 127, 129, 130, 142, 152
Honnell, Steve, 129
Hopkirk, Paddy, 27
Hot Rod, 12, 37, 48, 138, 147, 152, 170
Hotton, Andy, 23, 52, 90
Hudson models, 7, 35
Huggins, Jerry, 148
Hurst, Vanda, 40
Hurst shifter, 132, 163, 171
Hurtubise, Jim, 113
Hutcherson, Dick, 57, 65, 70, 103-105

Iacocca, Lido A. "Lee," 44, 129, 136
Indianapolis 500 race, 86, 135
Indianapolis Raceway Park, 46, 60, 64, 65, 88, 92, 135, 136, 142, 143, 153, 161
International Motor Contest Association (IMCA), 57, 65
International Motor Sports Association, 175
Isaac, Bobby, 63, 122, 130
Iskendarian camshaft, 21
Itala car, 6

Jaguar models, 20, 21
Jarman, Trant, 25
Jarrett, Ned, 45, 61, 66, 68, 70, 103
Jeffords, Jim, 153
Jenkins, Ab, 53
Johns, Bobby, 39
Johnson, Bob, 136, 144
Johnson, Junior, 39, 57-58, 63, 66, 70, 103-104, 112, 113, 122, 130
Jones, Parnelli, 57, 58-59, 62, 64, 65, 70, 104, 105-106, 113, 121, 130, 135, 149, 160, 161, 165-166

Joniec, Al, 153
Jopp, Peter, 25

Kalitta, Connie, 34
Kelsey-Hayes wheels, 16, 17
Kempton, Dave, 153
Kneifel, Chris, 175
Knudsen, Semon E. "Bunkie," 121, 129, 160
Koni shocks, 143, 145
Kuhn, Burl, 88
Kulich, Frank, 6
Kwech, Horst, 161

Laguna Seca race, 160, 165
Lamborghini Marzal show car, 126
Landy, Dick, 153
Lawton, Bill, 92, 142-143, 145
Leal, Butch, 65, 92
Le Mans race, 82
Leslie, Ed, 149
Lidden, Henry, 23
Lindamood, Roger, 92
Lions Drag Strip, 29
Ljungfeldt, Bo, 25, 27, 28
Lorenzen, Fred, 36, 39, 40, 44, 45, 48, 49-50, 53, 57, 58, 62, 64, 65, 66, 70, 74, 75, 100, 104, 105-106
Los Angeles County Fairgrounds, 44, 142, 143
Lotus Cars, Limited, 135
Lotus models, 135
Lucas lights, 154
Lund, DeWayne "Tiny," 57-58, 113, 161

Machete show car, 108
Maggiacomo, Jocko, 22-23
Magic Cruiser show car, 73
Manney, Henry III, 26
Marchal lights, 154
Martin, Ed, 60
Maserati, models, 57
Matthews, Banjo, 40, 44, 74, 106, 123
Mauro, Johnny, 20
McCahill, Tom, 37, 137
McCluggage, Denise, 21
McCluskey, Roger, 123, 130
McComb, John, 144
McCulloch supercharger, 9, 11, 35
McDonald, Dave, 142
McKenzie, Margaret, 25
McLaren, Bruce, 82
McLaughlin, Matthew S., 130
McLellan Brothers, 31
McNeely, Tom, 31
Mechanix Illustrated, 37, 137
Mexican Road Race, 8
Meyer, Louis, 135
Michigan International Speedway, 130, 132, 161
Mid-American Raceway, 144
Midwest Association for Race Cars (MARC), 40
Miles, Ken, 82
Miller, Akton, 51, 59
Miller 200 race, 123
Milwaukee racetrack, 123
Minter, Milt, 165
Mission Bell 250 race, 149, 165
Monte Carlo Rally, 25
Montgomery, George, 136, 150
Moody, Ralph, 9, 53
Moore, Bud, 58, 62, 64, 70, 75, 100, 104, 129, 132, 149, 161, 165, 166
Moore Engineering, 160, 165
Motor Life, 15, 22, 37
Motor State 400 race, 130
Motor Trend, 10, 11, 12, 15, 17, 19, 48, 85, 116, 128, 129, 149
Motor Trend 500 race, 55, 62, 70, 104, 113, 121, 130
Motschenbacher, Lothar, 153
Mt. Tremblant Circuit, 165
Musgrave, Elmer, 58
Mustang I show car, 137, 138
Mustang II show car, 139

National 400 race, 40, 58
National 500 race, 106, 130
National Association for Stock Car Automobile Racing (NASCAR), 8, 9, 13, 14, 15, 18, 22, 23, 32, 34, 36, 37, 39, 40, 42, 44, 45, 46, 48-53, 55, 57, 58, 61-66, 68, 70, 74, 75, 88, 92, 93, 99, 100, 102, 104, 105, 112, 113, 114, 118, 119-121, 123, 127-130, 132, 133, 142, 148, 149, 150, 157, 161, 166, 167
National Council of Mustang Clubs, 161
National Hot Rod Association (NHRA), 17, 28, 42, 45, 46, 52, 53, 59, 60, 61, 65, 70, 73, 88, 90, 91, 92, 93, 99, 106, 135, 136, 142, 143, 145, 149, 150, 153, 161
Naughton, John, 111
Nelson, Norm, 39, 40, 41, 50, 51, 70, 123, 130
Newsweek, 138
Nichels, Ray, 40, 45, 51, 64, 106, 123
Nicholson, "Dyno" Don, 28, 44, 52
North Carolina Motor Speedway, 75
Northwestern Ford, 110

Offenhauser, car, 135
Oldfield, Barney, 6
Oldsmobile models, 7, 8, 24, 35, 98, 107, 128
Oliver, Jackie, 166
Ongais, Danny, 160
Ott, David, 145
Ottum, Bob, 160
Owens, Cotton, 14, 37, 39, 40, 106

Pabst, David Jr., 145
Pacific International Raceway, 149
Panch, Marvin, 21, 22, 44-45, 57-58, 62, 63, 64, 66, 68, 70
Pardue, Jimmy, 63
Parsons, Benny, 110, 114
Pascal, Jim, 64
Paxton supercharger, 20, 145
Payne, Randy, 123
Pearson, David, 104, 106, 113, 120, 121, 122, 123, 127, 130, 132, 133, 149
Pellegrini, Ron, 142
Penske, Roger, 149, 153, 161, 165, 166
Performance Associates, 144-145
Perry, Harry, 39
Petty, Lee, 14, 39
Petty, Richard, 14, 39, 49, 63, 64, 70, 75, 105, 113, 114, 121, 123, 130
Pike, Don, 144
Pike's Peak Hill Climb, 45, 51, 59, 65, 117, 123
Pinta, David, 31
Pistone, Tom, 14, 36
Platt, Hubert, 118, 123, 145, 153
Plymouth models, 12, 19, 24, 70, 74, 98, 113, 114, 116, 117, 121, 129, 135, 138, 153, 160, 163, 165
Pollard, Art, 78
Pontiac models, 24, 47, 49, 61, 88, 98, 107, 110, 117, 128, 146, 160, 165
Posey, Sam, 161, 165
Proctor, Peter, 139, 142
Proffitt, Hayden, 46, 52
Prudhomme, Don, 34

Rahm, Jim, 39
Rambler models, 19, 20, 24, 83, 137
Rathman, Jim, 41
R. C. Industries, 59, 90
Rebel 300 race, 40, 45, 48, 49, 50, 58, 64, 130
Rebel 400 race, 122
Reed, Jim, 21
R. J. Reynolds Tobacco Company, 18
Revson, Peter, 149, 153, 161, 166
Rice, Tommy, 20
Richter, Len, 142-143
Rindt, Jochen, 144
Ritchey, Les, 44, 59, 92, 143, 145
Riverside Golden State 400, 58
Riverside International Raceway, 56, 57, 70, 74, 113, 121, 144, 149, 166
Road America racetrack, 136, 145, 165, 175
Road & Track, 23, 26, 138
Roberts, Fireball, 10, 14, 21, 39, 40, 44, 49, 58, 62, 63
Robinson, Pete, 34
Rodriguez, Ricardo, 21
Ronda, Gaspar "Gas," 59, 65, 90, 92, 142-143, 153
Rutherford, Johnny, 57
Ruttman, Troy, 51, 57

Saab models, 25
Sachs, Eddie, 45, 142
Sanders, Frank, 44
Savage, Swede, 161
Schmitt, Mike, 61, 73, 145
Schumann, Laverne, 45
Scott, W. B., 6
Seattle Automotive Club, 6
Sebring race, 82
Shawmut car, 6
Shelby, Carroll, 23, 82, 136, 140, 142, 143, 153, 154, 161
Shelby Automotive, 111
Shell 4000 Trans Canada Rally, 26
Shellenberger, Bud, 70
Shrewsberry, Bill, 28, 59
Smith, A. O., 154
Smith, C. J., 6
Smith, Jack, 39, 44, 49
Snow, Gene, 28
Southern 500 race, 8, 45, 49, 58, 66, 75, 100, 114, 122
Speed & Custom, 61
Springnationals, 70, 73
Sports Car Club of America (SCCA), 110, 136, 143, 144, 145, 146, 150, 161, 165, 175
Sports Car Graphic, 21, 144, 160, 171
Sports Car Illustrated, 21
Stacy, Nelson, 40, 44, 45, 48, 49, 50
Stardust International Raceway, 149
Strickler, Dave, 29, 52, 88
Stroppe, Bill, 13, 21, 22, 56, 57, 58, 59, 62, 64, 65, 106, 113, 117, 123
Stroppe and Associates, Bill, 20
Studebaker models, 19, 21, 24, 45, 137
Sudderth, Johnny, 39
Sullivan, Don, 36
Sutkus, Rich, 15

Talladega 500 race, 123
Tasca, Bob, 44, 60, 88, 90, 91
Tasca Ford, 143, 147
Teague, Marshall, 7
Team Starfish, 144
Terry, Ed, 106, 123
The Wood Brothers, 37, 44, 45, 49, 57-58, 62, 68, 74, 104, 105, 112, 113, 120, 122, 130, 132, 133
Thomas, Bill, 52
Thomas, Herb, 8, 52
Thompson, Jerry, 166
Thompson, Mickey, 52, 92, 160
Thompson, Speedy, 40
Thornton, Jim, 92
Timanus, John, 144
Titus, Jerry, 143, 144, 149, 153
Tour de France, 139, 142
Trainor, Bill, 40
Trickle, Dick, 130, 166
Tullius, Bob, 144, 149
Turner, Curtis, 8, 14, 21, 38, 39, 44, 45, 51, 59, 74, 100
Turner, Don, 46

United States Auto Club (USAC), 9, 15, 36, 39, 41, 42, 45, 46, 50, 51, 53, 55, 56, 58, 62, 64, 65, 70, 74, 78, 88, 105-106, 110, 112, 116-118, 123, 128, 130, 160
Unser, Al, 113, 130
Unser, Bobby, 117, 123
Unser, Louis, 45

Van Beuren, Fred, 149
Virginia International Raceway, 144
Volkswagen models, 20, 136
Volvo models, 20

Wade, Billy, 62, 64
Ward, Rodger, 51, 58
Weatherly, Joe, 8, 14, 21, 40, 44, 49, 51, 58, 62
Weber carburetor, 70, 143
Welborn, Bob, 14
Weld, Skip, 28
White, Don, 42, 45, 51, 56, 62, 70, 100, 106
White, Rex, 40, 58
Wilson, Fritz, 14
Wilt the Stilt, 37
Winfield, Gene, 95
Winternationals, 59, 91, 142, 145, 150-151
Winton, Alexander, 6
Winston, 18
Wisconsin State Fair Park Speedway, 15, 104
Wood, Dick, 123
World 600 race, 49, 58, 63, 122, 130

Yaeger, Tom, 144
Yankee 300 race, 64, 65
Yarborough, Cale, 64, 74, 104, 105, 112, 113-114, 120, 122, 126, 130
Yarbrough, LeeRoy, 112, 122, 130
Yunick, Smokey, 8, 39, 44, 49, 57, 121

Zecol-Lubaid chemicals, 37, 40, 41, 45, 51, 56, 58, 70